SECRET LIFE OF THE CITY

D1337515

HANNA BJØRGAAS

Translated by
MATT BAGGULEY

Secret
Life
of the
City

How Nature Thrives
in the Urban Wild

GREYSTONE BOOKS
Vancouver/Berkeley/London

First published in English by Greystone Books in 2023
Originally published in Norwegian as *Byens himmelige liv: Historier fra den urbane villmarka*, copyright © 2021 by Cappelen Damm, Oslo
English translation copyright © 2023 by Matt Bagguley

23 24 25 26 27   5 4 3 2 1

All rights reserved. No part of this book may be reproduced, stored in a retrieval system or transmitted, in any form or by any means, without the prior written consent of the publisher or a license from The Canadian Copyright Licensing Agency (Access Copyright). For a copyright license, visit accesscopyright.ca or call toll free to 1-800-893-5777.

Greystone Books Ltd.
greystonebooks.com

Cataloguing data available from Library and Archives Canada
ISBN 978-1-77164-935-3 (cloth)
ISBN 978-1-77164-936-0 (epub)

Copy editing by Lucy Kenward
Proofreading by Alison Strobel
Jacket and text design by Fiona Siu
Jacket illustration [or photograph] by [TO COME]
Interior illustrations by Ragna Misvær Grønstad, except the spectrograms on pages 60, 61, and 62 by the author. The illustration on page 124 is based on a photo by Katelyn Solbakk, with her permission.

Greystone Books thanks the Canada Council for the Arts, the British Columbia Arts Council, the Province of British Columbia through the Book Publishing Tax Credit, and the Government of Canada for supporting our publishing activities.

This translation has been published with the financial support of NORLA.

Greystone Books gratefully acknowledges the xʷməθkʷəy̓əm (Musqueam), Sḵwx̱wú7mesh (Squamish), and səlílwətaʔɬ (Tsleil-Waututh) peoples on whose land our Vancouver head office is located.

Contents

INTRODUCTION
An Old Acquaintance   *1*

JANUARY
The Urbanite's Reflection   *9*

MARCH
The Night Singer   *48*

APRIL
The War on Ants   *65*

JUNE
The Seagull Paradox   *90*

JULY
The Ghosts of the City   *113*

AUGUST
Stories From the Underground   *124*

OCTOBER
Marvels of the Darkness   *167*

NOVEMBER
The Written Language of the City   *194*

DECEMBER
The Near and the Dear   *208*

Thank You   *225*

Appendix: Author's Comments on the Text   *227*

Notes   *229*

An Old Acquaintance

IT BEGAN WITH a yellow-orange spot. I was standing opposite the penguin colony, on a rocky outcrop jutting up from the snow a few hundred yards above the shore. I looked a bit like a penguin myself. The multiple layers of clothing under my thick down jacket made my arms stick straight out from my body. That summer I was working as a guide for Antarctic cruise passengers, and my job that day was to make sure the tourists didn't stray from the designated paths and to answer any penguin-related questions.

A gray-haired lady, wearing one of the cruise operator's red jackets, soon made a beeline for me. As she made her way across the snowfield, she paused to allow a couple of lively penguins to run past on their way to the sea, then caught her breath while she watched them dive in. When she reached the place where I was standing at the top of the hill, she said, in English but with a German accent, that she wanted to show me a photo she had just taken.

I raised my sunglasses from my face while shielding my eyes from the glare of the sun. The photo she showed me didn't have any penguins in it. It was a close-up of some round, orange-colored shapes. It looked like a collection of glazed bowls set on a yellow-orange carpet of solidified lava. "Isn't it beautiful? I saw it right down there," she said, pointing toward the penguin colony. "What is it?"

Although they were quite a way off, I could see that the stones she was pointing at had an orange tint to them. In fact, the whole area around the penguin colony looked faintly orange.

I looked at the photo again. Something about it was strangely familiar. Then I realized that what she had photographed was a lichen belonging to the *Xanthoria* genus. I had seen this yellow-orange lichen many times before in more familiar surroundings—on trees, rocks, and concrete—and I was sure I could remember something similar growing on a tree in my own backyard. Now the two of us stood marveling at the lichen's impressive structure as though we were looking at it for the first time. It was beautiful.

AND THAT'S HOW my story about urban nature began one glorious summer day in December 2017, in as unlikely a location as Deception Island, just off the coast of the

Antarctic Peninsula. The island, essentially a crescent-shaped mountain range with a lagoon at its center, is so named because it is in fact the peak of an ancient volcano. A narrow gap between the mountains allows ships to sail into the extinct volcano's crater, which is now filled with icy Antarctic seawater.

The lagoon has been the setting of two of humanity's most determined attempts to wipe out parts of the planet's wildlife. At first the island was a base for hunting fur seals, then whales, in an era that almost destroyed the populations of both. Today, cruise ships, not fishing boats, keep the island's beaches occupied with people.

On the beach that day were two enormous seals, drowsily waving their flippers in the air. The tourists stood in a semicircle around them and took photos. Where I was standing, a few hundred yards up from the beach, hundreds of chinstrap penguins huddled in their nests among the black volcanic rocks. The snowfield that sloped down from the colony to the sea was like a penguin motorway. From my vantage point beside it, I could watch them slide on their bellies to the edge of the steep snowfield and jump clumsily into the sea, where they became elegant underwater torpedoes.

THE TOURISTS VISITING the island that December day had paid through the nose to see penguins, seals, and spectacular views. And they got their money's worth. Cold, nutrient-rich water from lower latitudes, pushed south by ocean currents in layers far below the surface, is forced toward the surface when it meets the Antarctic continental shelf. The seawater in this region is so rich in nutrients that huge blooms of algae can form when spring and light come to Antarctica. And the enormous number of calories in that algae provides sustenance all the way up the food chain, from minute zooplankton to seals, whales, and seabirds.

A FEW HUNDRED YARDS from shore was the gigantic cruise ship we were living on. It looked like a LEGO boat when set against the vast and magnificent scenery which, on our voyage south, had offered us one spectacular view after the next.

Here on the rocky outcrop by the penguin colony, the German lady and I looked out over a large glacier. At random intervals the glacier would calve, sending pillars of ice crashing down into the ocean. Large and small icebergs floated in the water like sequins scattered on a dark velvet sheet, and each new icefall caused the sequin carpet to dance upon the waves. The falling ice blocks looked small from where we were standing, but our view was misleading. I knew those blocks could be as big as houses, or apartment buildings, and when they separated from the glacier, they created waves large enough to capsize the many small tourist boats chugging noisily around the icebergs.

It is normal for the ice to calve during the summer, but the sight of those icebergs in the water was a reminder that most glaciers on the Antarctic Peninsula are melting faster than usual.[1,2] I thought about how our carbon-dioxide emissions were contributing to this faster rate of warming.

Most of us on the ship had also flown vast distances to get to Antarctica. And for every hour we sailed at cruising speed, our ship burned over 265 gallons of diesel.[3]

The idea of the landscape before us being untouched by humans was an illusion. Twenty-five years ago, Deception Island had perhaps twice as many penguins as it does today.[4,5] And one of the reasons for this decline is climate change.[6]

We guides had been told by the tour operator that this strange and beautiful landscape would touch the people seeing it, that these voyages create "ambassadors for nature." We were told that standing beside a massive glacier with near-luminous blue ice gleaming deep within its cracks, and visiting penguin colonies that have been stable for tens of thousands of years, would stir something within us. The idea was that these experiences would make it more important for the tourists, for us all, to make good environmental decisions back home. Our job as guides was to make sure the tourists had memorable encounters with nature, that nature made an impact on them. It should have been easy.

But something wasn't right. Although some of these people were fulfilling a lifelong dream, others seemed quite indifferent to what they were seeing. They came, took their photos, wandered around a bit, and left.

I felt like I understood why. These trips ashore were subject to strict rules and regulations. No one was allowed to walk around and explore and discover things on their own. To ensure efficiency and safety in this alien landscape, the excursions had to involve as little unpredictability as possible. The list of what tourists could see was specific and came with a price tag. It was an Antarctic adventure with a carefully edited script. On the other hand, there was no lack of beauty. The tour operators are selling the status that comes from posting a photo of Antarctic glaciers on Facebook or Instagram. Instead of whale blubber and sealskin, what's being consumed now is beautiful scenery.

I came to realize that I found the herds of tourists wearing borrowed red jackets and ill-fitting hiking boots—all of them attempting to take selfies with the penguins without getting other tourists in the shot—slightly ridiculous. It was hard to admit, but it was true.

WHAT ABOUT ME? I too had been travel bragging on social media, to overwhelming numbers of likes. But my degree in biology didn't render my trip more valuable or less harmful to the vulnerable Antarctic environment. I had also been lured there by the desire to see things that few others have seen.[7] Standing there with the Antarctic glaciers and penguins before me, and with an iPhone photo of an orange lichen in my hand, it dawned on me that I knew more about Antarctica's lichens than I did about the lichens in my own backyard. It was food for thought.

I HAVE SPENT significantly more time in the city than in the wilderness beyond it. It's where I've always lived. I grew up in Trondheim, a medium-sized city in the middle of Norway, in a residential area near the center. Several coincidences, along with a vague interest in nature, had led me in the direction of biology and botanical studies. But I hadn't found those studies particularly easy. Remembering the Latin names did not come naturally, as it clearly did for some of my fellow students who were hobby-biologists while growing up. I felt overwhelmed by the hundreds of species that make up an ecosystem, and they all seemed alike.

What saved me during my student years was the realization that I could fall in love with the details. To train my eye on a particular thread in the tangled web of different species felt great. It was like learning a craft. But I felt something else in addition to the sense of mastery. The process of scrutinizing plant stems with a loupe, searching for ligules or forked stem-hairs, hidden details that distinguished this particular plant from other species, brought me a sense of peace I had never experienced before. Perhaps driven by some latent passion for collecting, I could walk around for days in the woods, in the mountains, or by the shore, trying to find out what was living there. Recognizing a species I had seen before was like meeting an old friend.

The world of biology opened the door to field courses, excursions, and jobs. Eventually biology also took me to a couple of the world's most pristine natural areas, which I explored—binoculars and loupe at the ready—while looking, learning, and understanding.

But when each trip ended, I took off my binoculars and loupe, packed them in my rucksack, and went home to the city—which was now Oslo, to the orange lichens and crows and dandelions.

BY THE TIME I returned from Antarctica, it was January. All the snow had melted and the city looked drab. My adventure was over, and there were no more trips on the horizon. Far from it in fact. For various reasons, I was going to have to stay in the city for some time. And that bothered me. City life is okay when I'm feeling good. But when things aren't going so well, I often find that it helps to be in the forest, or the mountains, where no more is demanded of me than my presence. And when things get really bad, all I want to do is stroll around, without direction or purpose, in places where there are trees and birds and lichens to look at. And, in the vacuum of returning from a long trip, the restlessness stirring inside me was telling me I needed to get out of town.

But as things were, I had to make do with staying put, in a cramped apartment overlooking one of Oslo's busiest roads.

On my first day back in Oslo, I had no plans other than getting over the jet lag. I walked around the streets trying to unwind. The pigeons outside the subway station were pecking away at what looked like a squashed pastry. Swarms of sparrows chirped noisily from the ornamental bushes. Everything around me seemed to be shaped by humans. The wonderfully complex nature I had just returned from had been reduced to this: sparrows, pigeons, hard surfaces, and right angles. A destitute landscape. I turned and walked home.

When I arrived back at my apartment, I noticed something. Next to the bike rack in the backyard stood a mountain ash from which, just a day earlier, I'd hung a couple of suet balls for the birds. But all that remained of the suet balls were the green plastic nets, now dangling mournfully from their respective cotton strings.

The Urbanite's Reflection

How many times a day do you spy on a crow?
How many times does a crow spy on YOU?

I T TOOK A ROBBERY for me to understand the potential in those black crow eyes.

The suet balls hanging from the tree in the backyard were the stolen items; but until then the small, acrobatic blue tits had had a virtual monopoly over them. The sparrows stayed right where they were in the bushes. The magpies had a go, but since they're not built for hanging upside down, they flew off in shame after a few bungled attempts.

The day after I hung up the suet balls, they were gone. All that remained were the two green nets wafting in the breeze. And there wasn't a blue tit in sight.

Slightly puzzled, I went back into my apartment, fetched a pair of scissors and two new suet balls, and went back out to the tree. The mesh bags had been torn open, as though they had been slashed with a knife. No blue tit could have done that. I hung up the new food balls, and kept an eye on the tree through the kitchen window.

The next morning I saw a crow sitting on the branch the suet balls were hanging from. I hid behind the kitchen curtain to avoid being seen. The bird cocked its head toward the ball dangling a few inches below, then it bent down, grabbed the string with its beak, and hoisted the suet ball up a bit. The crow teetered for a few seconds with the ball still dangling well below the branch. Then it fumbled about trying to step on the white cotton string, before finally holding the string down with its foot. The bird then opened its beak and shook its head to free itself from the dry cotton string. The suet ball was a couple of inches closer to the bird.

The crow teetered a bit more, struggling with the heavy load, before once again stepping on the rope and letting go with its beak. The ball was now just below the branch. Next, the bird bent down carefully, gripped the top of the net, and with a jerk of its head, pulled up the suet ball like an angler flipping their catch into a boat. The momentum caused the crow to stumble backward, but with a few thrashing wing strokes it righted itself. And there it stood, with my suet ball safely in its care.

It struck me that if I ever had to pull a rope with just one arm, that's exactly how I would do it.

If you have ever watched a crow picking apart a garbage bag, you won't need me to explain how quickly this crow tore a hole in the mesh around the suet, which, once released from the net, promptly fell to the ground. And as I watched the crow fly away, heavily laden, I thought about how determined it had been. Nothing about what the crow had done seemed random. It had a plan. Had it learned the trick from someone, or come up with this solution on its own?

At the time I had quite an aversion to crows. I'd once seen them snatching baby eider ducks that were bobbing about in a row in the sea like little balls of wool. The mother duck was at the front of the line and had no chance of saving her ducklings from the crows. I also found it almost perverse how well these wild animals had adapted to living off our garbage. I saw crows as unwanted, unworthy city dwellers, unlike other, more likeable city birds such as the aforementioned blue tits. If I went out with a bag of stale bread to feed the ducks by the river, I would try to throw the food toward the ducks specifically so that the crows didn't get it.

But watching the crow solve the problem with the suet ball had been fascinating. I had seen plenty of crows in my life, but they always seemed to be operating in my peripheral vision. Had I *really* seen one? What did it look like? I sat down at the kitchen table and tried to draw the crow I had been spying on. But I just couldn't come up with a good picture. My pencil scrawled a dark and shapeless creature. The only recognizable feature was its beak, a sharp, scythe-like weapon. It vaguely reminded me of Pesta, the hunched and hooded woman who personified the Black Death in the drawings of Norwegian artist Theodor Kittelsen.

But it was while watching the crow haul up the suet ball, one beak-length at a time, that I had noticed a certain glint in its eyes. There was something about this creature that intrigued me.

JANUARY TURNED OUT to be a good time for surveilling crows.

When fall comes and darkness looms over the afternoons, crows begin to cluster—leaving their territories, dissolving their relationships, and becoming part of a large flock. These winter flocks can be enormous.[1]

One of these flocks ruled the palace gardens in Oslo where I went one afternoon. A fresh snowfall earlier that day had covered the old, compacted snow like a blanket of cotton wool. It was a totally calm day. The trees looked like black pillars against the pale blue sky. The wide slopes leading up to the palace, which normally bustled with tourists in the summer, were almost empty. But there were a few people here and there, walking or cycling purposefully through the park.

The crows also looked like they had something important to do that day. They hopped around in the snow or fluttered between the branches. One bird landed in a large birch tree, its weight sending fresh snow cascading silently to the ground. I sat down on a bench a few feet away and took out my binoculars.

I had discovered that spying on crows was quite hard, but not because of any technical difficulty. It was the opposite. Crows were everywhere when I first started noticing them, and they came so close there was almost no point in having binoculars. The problem was me. I was so used to walking past them that it required a huge effort to actually stop. Even having decided to look at the birds, my first impulse was to go and spy on the people walking along Oslo's main street—Karl Johans gate—or check my phone for email instead.

But I had decided to spend half an hour watching crows, and so I pointed my binoculars up at the birch tree. A crow squinted back at me and gave me a few moments to study it. Its furrowed brow sat low over its eyes, and its wide beak gave the impression of a broad gape with downcast corners not entirely unlike an elderly Mick Jagger's mouth. The crow didn't seem to like being looked at. It became quite vigilant when I raised my binoculars, as if it knew I was getting right up close to it, if only with my eyes.

The crow's large head was bent forward, an extension of its powerful neck. It seemed front heavy. Its head, chest, and wings were all covered in black feathers that were curiously both matte and shiny. The rest of its body was covered in a matte, slate-gray plumage. Up close, I could see that these feathers went all the way up to the underside of its beak, giving it a kind of downy chin. The crow stared right at me. Admittedly a crow's pupils, irises, and the feathers around its eyes are totally black, like a person wearing dark sunglasses, so it was hard to say exactly where it was looking.

It was quiet in the park. The roar of traffic around the nearby National Theater had been reduced to a faraway hum, deadened by the snow. Now and then the silence was interrupted by a couple of muffled crow calls, *caw! caw!*, which sounded like parts of the birds' daily conversation. A group of pigeons pecked around in the snow beside a garbage bin. A cyclist exhaling foggy breaths rode up the wide slope in front of the palace, the snow creaking under her tires. She rode past the pigeons, then turned onto a trail and vanished among the trees.

A YOUNG MAN wearing a suit and overcoat had come up the hill and was walking toward the larch tree where the crow was sitting. He was holding something wrapped in white paper with the word "SUBWAY" written across it. His dress shoes looked unsuitable for the snow. The crow in the tree nodded its head. It blinked under its black hood.

The man then slipped on the icy path and, with his arms flailing, let go of what was in his hand. When he regained his balance, he paused for a second and continued walking, leaving the half-eaten sandwich he had dropped lying in the snow.

The crow immediately lost interest in me. As soon as the man was out of sight, the bird flew down from the tree and landed on the ground beside the man's abandoned lunch. It pecked at the sandwich from different angles, then tried to pick it up with its beak. But the two halves of the bread just fell apart, exposing the filling inside. The crow then looked around, picked up the half with cheese and ham stuck to it, and quickly bounded off toward some nearby bushes. The bird struggled through the stiff branches until it reached the middle of the bush and left the bread by the roots. Then, using its beak as a shovel, it buried the sandwich in the snow.

The crow reversed out of the bush, looked around again, as if checking to make sure no one had seen what had happened, and quickly went back for the other half of the sandwich. It then picked up the remaining food in its beak and, with its heavy load, flew away.

The Crow Watcher

"I've been awake since five o'clock," said Geir Sonerud, his face stretching as he yawned. He spoke in a calm East Norwegian dialect that made me wish I was making a radio show instead of writing a book. "It used to be bird-watching that dragged me out of bed; now, it's my age."

There we sat, cups of tea in hand, at the researcher's kitchen table at his home in west Oslo. He looked in surprisingly good shape for someone almost seventy years old. Other than waking him up early, age didn't seem to have affected Geir to any noticeable degree. It wasn't clear if the spring in his step as he strode across the living room was due to the antioxidants in the green tea or the bowl of nuts and dark chocolates on the table. I ate quite a lot of the chocolate anyway, just to be sure.

Geir is a biologist and a professional crow watcher. He told me that when he began studying biology, urban nature was not considered "nature" among his contemporaries. Nature was something that went on beyond the city limits, and only the most pristine nature was considered worth studying. In keeping with the spirit of the times, it wasn't really crows, but the elegant birds of prey that caught his attention when he was a young boy.

"I'm really embarrassed about how blind I was to urban nature," he laughed.

I had told Geir about my experience in the palace gardens, and the incident with the suet ball. Both stories had made him wildly excited.

He told me that he and his colleagues had set up an experiment which was actually aimed at northern goshawks. They had placed three brown chicken eggs in a fake nest in a small area of woodland.[2] Geir had then sat quietly on the outskirts of the woods, camouflaged and equipped with a telescope. But the goshawks didn't show up, and Geir started getting bored. Until something else captured his attention. A crow had found the nest containing the eggs. Crows love eggs.

Three chicken eggs are too much for one crow to eat at once. So, what would it do with the eggs it couldn't eat right away? In fact, the crow didn't waste time eating *any* of the eggs. It looked around. Then it picked up one of the eggs in its beak and hurried off to a nearby spot where it put the egg on the ground and hastily covered it with moss and lichen. The crow rushed back to the nest and picked up another egg, and this time it flew a little bit farther away.

With two of the eggs properly hidden, the crow had plenty of time to consume the third and last egg at its leisure.

"The crow won't waste time eating the first egg; it'll make sure it gets as many eggs to safety as possible," said Geir.

He explained that this crow had hidden the eggs based on stringent mathematical logic. Every second counts for the crow, which is surrounded by rivals on the food front. It must find the best compromise between protecting the rest of the eggs in the nest and hiding each egg well enough. While the first egg is being hidden, the nest is still full of valuables, so the crow will hurry up and return as quickly as possible to protect the nest from competitors. The more eggs the crow has already hidden, the fewer valuables there are left in the nest. This makes it more prudent for the crow to spend time hiding the remaining eggs properly, instead of rushing back to the nest.

The crow can then eat the last egg, confident that the other eggs are safely hidden for a later date.

Geir repeated the experiment several times, and the result was always the same.

"Even though it is hungry, the crow thinks long term. It has self-discipline."

THIS STORY REMINDED ME of a study about humans and our ability to suppress our immediate desires. Forty test subjects were given the choice of getting something they like to eat—let's say chocolate M&M's—either two chocolate candies right away, or six if they waited one minute. All of the test subjects said they would be happy to wait one minute for six chocolate candies. Then the two chocolate candies were put in front of the test subjects, and they were given the option to wait one minute without eating them to get four more. But only one in five test subjects managed to wait the required time to triple the prize.[3] It would appear that the crow beats the human when it comes to self-control.

Geir told me that he was once sitting at the edge of the woods with his colleagues when they spotted a crow that had found the artificial nest. The bird was right in the middle of hiding an egg.

"The crow looked around like it was making sure no one had seen where it had hidden the egg," laughed Geir. "You could almost imagine it whistling innocently to itself. *Nothing going on here!*"

But then it stopped abruptly. The researchers couldn't understand what was happening until they glanced in the direction the crow had been looking. Sitting in a nearby tree was another member of the crow family, a Eurasian jay. And it was keeping a close eye on the crow.

The very next moment, the crow pecked a hole in the egg and gulped down its contents.

"Had the crow hidden the egg as planned, the jay would have taken the egg the moment the crow's back was turned. So once the jay had seen the hiding place, there was no point in continuing to hide the food! The best option was for the crow to just eat the egg right away."

"Do you think the crow would have reacted if, for example, a pigeon had seen the hidden egg?" I asked.

"No, I don't think so. I think the crow fully understands which of the other birds are smart and which ones are stupid," he replied.

And this is why I'll never underestimate a crow again. Geir's fellow researchers believe that crows have a theory of mind, an ability to understand another's behavior.[4]

To find out what birds understand about the intentions of others, and what insight they have into the mindset of other birds, researchers conducted a series of experiments on California scrub-jays. This species, *Aphelocoma californica*, is an omnivorous member of the crow family that lives in North America.[5]

The researchers found that scrub-jays that have committed acts of theft themselves understand that they, too, can be vulnerable to theft. If another bird sees where they are hiding food, they will move that food to a new hiding place.

However, naive scrub-jays that have never *themselves* stolen food from other birds are unable to imagine anyone else doing so. Even if they notice that another bird has seen where they have hidden food, they won't move the food and hide it elsewhere.

The researchers believe that these birds use their own experience to imagine what it is like to be in the other bird's place. In other words, it takes a thief to know one.

When you realize that someone wants something that's yours, the next step is to try to manipulate them into not robbing you. In a number of experiments with ravens, researchers have shown that ravens will often try to prevent other birds from seeing where they are hiding food. They will hide it in secret, behind an obstacle for example, or fly just far enough out of sight of the other birds. The raven might also delay hiding the food and wait for the other birds to become distracted, as humans might do with children when hiding Easter eggs.

Deception is also not unknown in the crow world. Ravens that are low in the flock hierarchy, usually the weaker birds, will easily lose the tussle for food if a fight breaks out. So, let's say that one such lesser raven is the only bird that knows the location of some hidden food. It would then be in the bird's interest to lure the other, stronger ravens in the wrong direction. The lesser raven will attract the attention of other ravens by pretending to have found food. The stronger ravens will then waste time searching in vain for food in that place, while the lesser raven sneaks off to where the food is genuinely hidden.[6,7]

Ravens are even known to make empty food caches. These caches trick the other birds into spending time looking, while buying the first raven more valuable time to eat the actual food.[8]

The ability of birds in the crow family to understand the mindset of other birds—what they think and understand— is a mental tool that few animals possess. Human children are often four or five years old before they develop the ability to grasp what others understand.[9,10]

What other secrets is this genus hiding?

All sorts of strange and unexpected things can emerge when you dig a little into the family history. And the crow family tree has already produced a few surprises. First, nothing about the crow's voice suggests that it is a singer. But though it lacks any obvious musical talent, the crow family, *Corvidae*, actually *is* part of the songbird family. They may not use it very often, but crows have a well-developed syrinx, or voice box.

Surprise number two: one of the crow family's closest relatives is the extravagant bird-of-paradise family.[11]

Both crows and birds-of-paradise are native to Papua New Guinea. But while the spectacularly colorful birds-of-paradise live only in Papua New Guinea, one of the least urbanized places on Earth, the more discreetly attired crow family has conquered the planet.

The large crow family is divided into several genera, one being the genus *Corvus*. This genus contains more than a third of the species in the crow family, all of which have *Corvus* as a prefix in their Latin names. The *Corvus* genus has succeeded in spreading over large parts of Earth's land surface, and to many different environments. It is quite comforting to know that almost anywhere you travel, you'll be under the watchful eye of a close relative of the crows you see back home.

Some birds in the crow genus thrive more than others. The common raven, *Corvus corax*, can survive in most of the world's environments, from the arctic tundra to the desert to the islands of the Pacific. The omnipresence of this charismatic creature might be one of the reasons why it plays a part in traditional stories in many places in the world. In Scandinavia, the Norse god Odin's delegates took the form of ravens. From their vantage points, Odin's very own winged messengers—Huginn and Muninn—brought him news from most corners of the world. However, ravens prefer to avoid cities and other densely populated areas. So, Odin would have missed out on a lot of useful city gossip had he trusted his ravens blindly. Had he sent out bird spies today, Odin may well have chosen another species.

When it comes to thought and memory (*Huginn* and *Muninn* in Old Norse), crows are on par with ravens. But while ravens will turn back at the city gates, crows will follow us right in. Members of the crow genus are among the very few species that have enjoyed the expansion of human settlements. Many crow species are thriving in habitats created by humans.[12] And we come closest to crows in the cities. In the countryside they are often far more shy.

However, in cities around the world, the word "crow" might refer to different species. The bird species commonly called "crow" in Norway and many parts of Europe is *Corvus corone cornix*: the hooded crow. Unlike its relative the all-black American crow, *Corvus brachyrhynchos*, the hooded crow is gray apart from its wings and the area covering its head and chest that can almost resemble a black hood.

Magpies are the crow's elegant cousin, from the same family but a different genus. With its metallic black-and-white tuxedo, its long tail and white breast, I'd be tempted to say it was beautiful had I not become so used to seeing magpies hopping about nearby.

Today's crows are descended from a "primordial crow" that moved out of the dense rain forest to more open and grassy steppe regions about twenty-eight million years ago.[13] When we apes climbed down from the trees and migrated to the savannas five to six million years ago, crows had been there for millions of years already. Like us, they often prefer more open areas with a light covering of trees to dense forest. From a crow's perspective, for example, a garden surrounded by large trees is ideal.

Gossip and Death

Back at Geir's house in Oslo, the bowl with healthy sweets was now empty. Geir apologized for not knowing where his wife had hidden the real deal, the sinfully sugary chocolates. After pouring more tea, he pointed out the window at one of the neighboring gardens.

Out there, on a hill overlooking Oslo Fjord, was a huge tree with light gray bark that was smooth as skin. Geir said the tree grew stiff red leaves in the spring. It was a copper beech tree. But now, in mid-January, it had a rounded silhouette formed by leafless branches. These branches extended from a low point on the trunk and stretched evenly toward the sky, making it a perfect climbing tree. It must have been visible from far away.

According to Geir, the copper beech is one of the city's regularly used pre-roosting trees. Every afternoon during the winter, he said, crows will come to this tree and chatter and make a lot of noise, as they have been doing for decades. He said there were lots of these pre-roosting trees in Oslo, in the cemeteries and in the park around the palace for example, and they were used so frequently you could mark them on a map.

"A crow will hang out there for a while, bickering and squawking, before flying off to a different pre-roosting tree. Sometimes it will return to the first tree, before making a final decision about where, and not least with whom, it will spend the night."

When darkness falls, the crows in the tree will calm down. Then, they will fly from the pre-roosting tree, quite silently, to the tree in which they have chosen to roost.

I have heard and read that strange things happen at these so-called crow meetings. They say that crows are put on trial there. Some people claim to have seen crows being expelled from the flock, or simply being pecked to death. A friend said she had seen a dead crow lying in a field, surrounded by a huge flock of crows. They weren't eating or fighting; they were just *there*, standing by their deceased cousin. She wondered if she had witnessed a crow trial and subsequent execution.

While these beliefs about crow trials and executions are anecdotal, a number of studies shed light on what might actually be going on at these crow gatherings.

To find out if crows are transferring information in these pre-roosting trees, Geir and his colleagues attached radio transmitters to crows in a rural area a few miles north of Oslo. They then placed large piles of offal in random spots near places they had seen the crows. Some of the radio-tagged crows found the offal and were allowed to eat their fill. When it was time to sleep and the birds left the offal, the researchers noted which trees the lucky crows spent the night in.

The next day, the researchers were monitoring the feeding spot, and noticed, unsurprisingly, that the radio-tagged crows that had found the offal the previous day returned to it. But that wasn't all: this time they brought more crows with them, including many that hadn't found food the previous day but had been in the same pre-roosting tree as those who did.[14,15] So they, too, got a share of the meat. According to Geir, there was no indication that the crows with prior knowledge of the food had tried to get rid of their less-fortunate colleagues. Quite the opposite, in fact.

"It is clear that when crows roost in this manner, there is some form of information exchange about who has found food that day," he says.

There may be a few reasons why crows tag along with the crows who have found food—odors might linger in their plumage, for example—but several studies of this crow genus indicate that Geir is right. Ravens in particular have been thoroughly studied. They, too, gather in pre-roosting trees, and researchers have observed that when ravens find food during the day, they can put on acrobatic aerial displays for the other ravens when they return to the pre-roosting tree.[16] They will roll and soar in the air in a way that seems like pure showing off. And the ravens doing the air show will be the same ravens leading the flock back to the feeding area the following day. In the morning, before leaving with the new members of their flock in tow, these ravens will screech and squawk at the roosting tree—as if reminding their fellow crows that it is time to leave.

Other studies have shown that ravens that have found a good place to eat may even deliberately look for the rest of the raven flock in order to show them the food's location.[17] What these individual ravens gain by sharing food has been studied using mathematics: experiments based on game theory have shown that the birds actually *do* gain something from alerting their colleagues—under certain conditions.[18]

When food is found in large quantities but is difficult to locate, it pays to share.

Geir explained that winter, and snow, are among the conditions that apply when it comes to crows. Snow will obscure whatever's lying on the ground and make food more difficult to find. The crows will then have to cooperate more. When food first appears—whether it be an open garbage container, a dead moose on the highway, or a bag of prawn shells that someone left out on the porch—there is often more than one single crow can eat. Sharing costs very little, and individual crows often depend on the crow community for their next meal.

The crows' pre-roosting spots remind me of après-ski parties, where skiers meet at a bar after a day on the slopes. Beer will be drunk, and important information shared. Where was the skiing good that day? Someone will have found great powdery snow on a slope with low avalanche risk and will be going back the next day. The others, who weren't as lucky in their hunt for perfect snow, will listen intently. Most people will share this information because they know that finding a good spot is based mainly on luck. The sharers will benefit later when others find good snow and they haven't been so lucky themselves.

In March and April, the supply of food becomes more reliable for birds and it makes sense for crows to disband, according to Geir. In the spring, there are always worms to be pecked out of the soil, or tourists to rob. If the crow's territory is good, it will manage on its own. Individual crows no longer need each other, so they will put the camaraderie aside and abandon the flock. The large aggregations break up, and the crows revert to living in pairs.

But it was January—time for living collectively—when I spoke with Geir. I asked him why some crows spend the night in one place while others go somewhere else. They all want to be where the food is, surely?

"Crows have different needs, just like humans. One of them might be really hungry, so it will join the adventurous ones and fly far away to a huge garbage dump the following day. Another crow might be feeling tired and need a rest, so it won't want to fly that far the next day. It will tag along with the crows going down the road instead."

Geir and I got ready to venture out into the January cold. While wrapping his scarf around his neck, Geir said that it was hard to explain crow behavior without taking into account some kind of language. "There has to be something. Nothing about the crow surprises me anymore."

If such a language exists, humans have borrowed a few words from it. The word "crow" itself originates from the Old English word *crawe*, which imitated the sound of a crow call.[19] Or perhaps crows gave the name to themselves. *Craa! Craawe! Craaawe!*

The species name *cornix* is also probably derived from the Proto-Indo-European word *kor-n*[20] or *k' orh-*, which, just like the English word, is an imitation of the crow's call. And the genus name *Corvus* means "raven" or "crow" in many Latin languages. This word is derived from the Proto-Indo-European word *kor-ou*, which is also an imitation of the crow's song.[21]

But do these sounds actually mean anything?

I opened the door and we walked out onto the steps of the house. The frigid air made the hairs in my nose ice up. A crow gave off a loud *crraw!* and then flew out of the beech tree.

Geir told me that large flocks of crows will gather in the beech tree. Sometimes, in broad daylight, hours before they would normally gather for pre-roosting, as many as fifty crows might sit there making a racket. He had seen crows fly long distances, often from a mile or more, to join the spectacle.

The reason for this gathering, he explained, is often because a goshawk, one of the crow's worst enemies, is nearby. But I wondered: Why would a crow fly into a danger zone? Wouldn't it be better to get out of there?

Geir explained that the northern goshawk is a master of the surprise attack when it comes to individual crows. Crows are therefore safer in a flock that has many pairs of eyes and an overview of the situation. Sitting amid a noisy flock of crows offers protection against being surprised when one least expects it.

Since the seventies, we've known that many of the sounds crows make can provoke specific reactions from other crows.[22] What Geir can hear when the crows sit together in the copper beech tree outside the house is, for example, a staccato "scolding," when the crows will berate the despicable hawk for daring to get too close. The noise frightens the hawk, according to Geir, and alerts other crows nearby. The intensity of the scolding varies according to the type of danger and how close it is.

What we humans perceive as a uniform-sounding *craw!* can vary in length, pitch, and tonal quality. Every crow has a unique voice, which other crows can recognize.[23,24] Recent research suggests that some crows may also be capable of distinguishing between human voices.[25]

Geir and I strolled down to the sea along several residential streets. The crows in the trees greeted us with an indifferent *craw!* I told Geir that I had been visiting the crows in the palace gardens quite regularly. I told him how I sometimes took a bag of bread with me, and that the first few times I did it, all the crows had looked identical to me. Even far more driven biologists find telling one crow from another difficult. On average there is a difference in size between the sexes, but large females can be larger than small males.

Eventually I did start to notice individual differences between the crows. Some were clearly bigger. Some were shy and kept their distance, and others would come right up to me, tilt their head, and boldly look up at my face, so close that I would almost back away myself. Some were particularly good at catching pieces of bread in the air. One would just stay on the ground and rarely flew more than a few feet perhaps due to an injured wing. But I was still far from being able to tell them apart.

But what about the crows? Was I imagining it, or were they approaching faster and faster when I arrived with my bag of bread? Were they recognizing me?

To answer this question, I have to make a leap in time and space, to the University of Washington, Seattle campus specifically. It is February 2006, a few days before Valentine's Day. Seattle is a city of parks and lawns and scattered trees. It is cold, and very few students are walking on the straight paths crisscrossing beneath the trees on campus. On this winter's day, the crows outnumber the students who notice a couple of figures crossing the frost-covered grass. Their facial features are coarse and distorted, with heavy eyebrows and flat noses. What are two Neanderthals doing out on the lawn?

The crows on campus have for many years been the closest research subjects of biologist John Marzluff, who studies crow behavior. Hidden under the caveman masks, he and his student are about to test how farsighted the crows are, and how good they are at remembering an enemy.

Over the next few hours, the two Neanderthals catch seven crows. They place identification bands around the crows' ankles and let the birds go. A flock of highly agitated crows circles overhead during and after the entire operation. They screech their hoarse-sounding *craaw!*, a sharp alarm signal, before diving toward the Neanderthals, just as they would when a fellow crow is being attacked by predators or is in some other kind of danger.

Once they have captured and released the seventh and final crow, the Neanderthals enter a building, where, out of sight of the crows, they remove their masks and become crow researchers again.

A few days later, John Marzluff takes a walk through the campus wearing his mask. He has done this before without the crows showing any interest in him. But today is different. The first crow he encounters squawks loudly when it sees him. It does not have a band around its ankle, so it can't be one of the crows the Neanderthals caught. Yet it follows him, cawing loudly and keeping a wary eye on him. Maybe it witnessed the offense the two masked researchers committed a few days earlier? The squawking attracts one of the other crows, this one with an ankle band which means it experienced capture firsthand. Both birds keep Marzluff under close surveillance on his tour of the campus.

Of the twenty crows he saw that day, three reacted to him.[26] A small percentage of the campus crows realized that this Neanderthal was not their friend.

For the next two-and-a-half years the crows are left in peace.

In the fall of 2008, John Marzluff puts his mask back on and goes for a walk under the trees outside the office. His presence goes down badly among the crows, to put it mildly. Most of them, perhaps two-thirds of the thirty-eight crows he encounters, squawk angrily at him,[27] which is a far higher number of crows than was originally caught. When other people wear the masks, they too cause havoc among the crows. But when Marzluff isn't wearing his mask, he is allowed to walk in peace. The crows are reacting to the Neanderthal mask.

Ten years after the initial capture of the seven crows, most of the crows that witnessed the incident have most likely died. Nevertheless, in 2016, when Marzluff took another walk around the campus wearing his mask, more crows squawked at him than ever before.[28,29]

What the researchers think is that when a crow squawks at Neanderthal John, other crows rush in, not only to help with the cause of the squawking, but also to see what the danger is. The next time they see the Neanderthal mask, they have learned to associate it with danger, and they sound the alarm even though they haven't experienced the original wrongdoings. When the crow-parents squawk at Neanderthal John, their chicks learn that this face belongs to someone who is hostile toward crows. This way, a creature's reputation for danger can be kept alive over generations.

But crows can also become friends with humans. There are stories in the media about crows bringing gifts to the people feeding them. Some crows have left colorful stones, plastic toys, and dead mice on the doormat of their most faithful allies.

Geir confessed he had spent years trying to become friends with crows. "I'm quite bitter about it," he laughed. "I've been feeding crows and magpies for years, leaving food out for them several times a week, but it hasn't made them any more trusting of me. They just squawk at me and fly off. I never get any presents either, although they do occasionally drop walnuts on my car."

I chose not to remind him that much of his career has been spent catching and tagging crows, clearly against their will.

IN SEATTLE, John Marzluff had shown that flocks of crows can develop a form of collective awareness. Knowledge about dangers or things that are good, about which animals are kind and which ones are hostile, can be shared and disseminated over time and from generation to generation. This knowledge transfer creates a tradition, a crow culture, within the flock. While we see just a mass of gray-black crows, these birds are able to recognize individual humans. Perhaps crows know more about us than we know about them.

WHAT SHOULD WE make of the anecdotes about crows gathering around other dead crows?

Kaeli Swift has taken the mask experiments one step further. Swift, who was one of John Marzluff's students, wanted to investigate the crows' relationship with death. She had masked students feed the crows peanuts, in the same designated places for several days, until the crows started waiting for the masked students to arrive.[30,31] The students then went to the feeding spot, still wearing the masks, only this time they had a dead crow with them. The crows already at the feeding spot, waiting for peanuts, immediately sounded the alarm. As with Geir's story involving the goshawk, more crows joined in, even though the only threat the masked students posed was the fact they were carrying a dead crow. The crows soon began to flock around the dead-crow carriers while squawking and making a commotion.

On each of the following days the mask-wearing students returned, with peanuts but *without* the dead crow. But the mood had changed among the crows. They took a long time to come for food. It was as though they were trying to avoid the place where the dead crow had been seen. The birds that did come squawked at the masked students, even though the students were no longer carrying a dead crow. Some crows went on the attack. It was obvious that the local crows had developed an aversion to both the feeding spot and the mask wearers.

In her dissertation, Swift calls the gathering of crows around the dead crow a funeral ritual. She believes that crows will use another crow's death to learn which places or creatures might be dangerous in order to avoid falling into the same trap themselves. The crows were like a study group learning about violent death.

Street-Smart Crows

We humans like to think of ourselves as the pinnacle of what evolution has managed to accomplish. In the search for characteristics that require complicated thought processes, researchers have looked mainly at the great apes; that is, humans and our close relatives. We have assumed that humans are quite alone in having a language, in understanding what others think, in making and using complex tools, and in gaining knowledge from others that we can store for later use.

In other animal groups, we have often looked for intelligence among those that visually resemble us. Owls, for example, are often considered the smartest members of the bird family, perhaps because they have flat faces, high foreheads, and eyes that face forward just like us? But owls are in fact fairly stupid (if you were to dunk an owl in water, you would see that its large head is predominantly made up of feathers). Crows, on the other hand, have low foreheads and eyes on either side of their head. Their appearance is quite different from ours. Yet in just the previous few weeks I had seen crows communicate and think long term and strategically. Could it be that humans have more in common with crows than I had previously thought?

"Mango" is a New Caledonian crow, an all-black bird with large glistening eyes and a short, triangular beak. Mango had been living a normal crow life in the jungle before being captured by curious scientists. It has been known for years that New Caledonian crows are extremely good at making and using tools. Prior to that, only humans and chimpanzees were thought to be capable of making tools out of several parts, without any guidance or fumbling. The ability to create complex tools to solve problems is very demanding on the brain and takes many years to develop in human children.

As part of an experiment, Mango and eight other New Caledonian crows were offered their favorite food (the researchers don't say what this was, but I imagine they used pieces of raw pork heart as they have done in other similar experiments). The problem for Mango and the other eight crows was that these tempting pieces of meat were in a cage and well out of reach of the birds. Outside the cage was a pile of sticks: some were large and hollow, and some were smaller in diameter. However, all of the sticks were too short to be used by the crows to fish the food out of the cage.

Although they had never seen these kinds of sticks before, Mango and three other crows successfully joined two of the sticks and managed to fish out the pieces of meat without much trouble.[32] When the researchers moved the food even farther into the cage, Mango managed to join *four* of the sticks. He therefore won both the food and the competition and made the short list of animals capable of constructing their own multipart tools.

If you bisect a human brain, you will see a thin gray layer that covers the underlying white matter. This gray matter consists mainly of cell bodies, and the connections between neighboring cells are made via synapses. In mammals, tasks involving complex cognition, such as decision-making and language, are primarily dealt with by an area of gray matter called the neocortex. The neocortex, the star of the story of human intelligence, comprises the largest portion of outer gray matter and is made up of six horizontal layers. In humans, these layers are only about an eighth of an inch thick all together, but they are folded into hills and valleys and therefore have a large surface area.[33] It is thought that this layered structure is important for the role that the neocortex plays.[34]

Originally, limitations in technology led researchers to conclude that a structure as complex as the neocortex existed only in mammals.[35] In the late 1800s, the German researcher Ludwig Edinger was interested in the structure of bird brains. He dyed pigeon brains using an early form of histological staining, a process which makes the structures of different types of tissue more visible, then cut the brains into thin slices and put them under the microscope so he could compare the brain structure of birds and mammals. Under Edinger's microscope, the bird brain looked almost homogeneous; it appeared to be devoid of layers, or any connections that bring different parts of the brain together. These structures, researchers believe, are among the things that make it possible to coordinate and process different sensory impressions and enable the more advanced mammals to reason and learn. Edinger therefore believed that birds have very limited cognitive abilities and that their complex behavior results from pure instinct.[36]

In Edinger's worldview, primates stood majestically alone at the top of the evolutionary ladder. However, the researchers had difficulty explaining how even a pigeon, with a brain the size of a shelled peanut, has fairly advanced cognitive skills. Pigeons can, for example, recognize faces and remember places, or be trained to distinguish between paintings by Monet and Picasso, even between works they have never seen before.[37]

In order to compare the intelligence of birds and mammals, one must know what one is measuring. Many animals can perform complex actions based on their genetic programming. Bees, for example, will dance to show each other where food has been located, and yet we don't ascribe them any great mental capabilities. Many species can also be trained to perform tasks or respond in a certain way to stimuli.

But some animals manage to solve *new* tasks that were not previously part of their repertoire. These tasks, therefore, are neither genetic nor learned. These animals think anew; they have a mental flexibility that enables them to behave cleverly when faced with totally new challenges. Many researchers consider this ability to be an important form of intelligence.[38]

Edinger's Austrian successor, the behavioral ecologist Konrad Lorenz, spent his whole life researching ravens, and also kept them as pets. Lorenz discovered that his ravens clearly behaved instinctively—they had a set of methods for dealing with the world—but this instinct could be modified through the experiences the birds were given. And he found that, unlike Pavlov's dogs, ravens learn in a similar way to humans: they are creative and flexible. This ability, according to Lorenz, indicated that a raven's brain had to be governed by more than instinct and acquired characteristics.[39]

Thanks to better dyes, color technology soon substantiated Lorenz's theories. Just as we can use chemicals to develop an image on photographic film, scientists were able to use these new chemicals to reveal hitherto unknown structures in cross-sections of a raven's brain. Slices of raven brain tissue stained using this new color technology showed that ravens' brains are anything but uniform.

Just like in mammalian brains, the white matter in bird brains is covered by a different kind of brain tissue.[40] However, whereas the gray matter in a mammalian brain is layered, like a sandwich, says crow researcher Nicola Clayton in an interview with *National Geographic*,[41] the white matter in a bird brain might instead be compared to a pizza. All the sandwich ingredients are there, the cheese and the tomato and the ham, but they are side by side rather than on top of each other. The pizza-like structures in the bird brain can work analogously to the structures found in parts of the mammalian brain's neocortex.[42] The bird brain also appears to have the capacity for interconnection and processing different sensory impressions.

Here lies maybe one of the keys to understanding the crows in the palace gardens. The crow's brain tissue is perhaps *similar enough* to the mammalian neocortex that it is possible for crows to have a cognitive way of dealing with reality that is comparable to the one humans have. Bird and mammalian brains have followed different evolutionary paths, but they have nevertheless ended up in a similar place.[43]

Neither mammals nor birds get their brain power for free. On the contrary, large brains come at a high price (anyone who has witnessed a *Homo sapiens* give birth will attest to that). A brain is heavy and takes up a lot of space.

It uses a disproportionate amount of the energy absorbed by the body. So why do some animal groups develop large brains while others don't? And why do crows and humans—perhaps two of the smartest animals—meet specifically in cities?

In the journal article "Intelligence in Corvids and Apes: A Case of Convergent Evolution?" the authors look at theories that aim to explain why primates developed large and costly brains. The authors believe that many of these same theories may also explain crows' large brain capacity.[44] For example, many primates live on plants and fruits, which grow in different places and mature at different times. The landscape offers a good source of food, but an individual must remember where and when this food can be found. So having a good memory is highly advantageous.

The article's authors believe the same argument can be applied to crows, which live predominantly on seasonal food. Walnuts, for example. Like the walnuts from the tree growing behind Geir's house.

"Typical September food," says Geir. "It's something city crows have added to their yearly calendar." Geir and I had just walked beneath a huge walnut tree with rough bark in a garden near his house. "In the fall, crows come from all over the city to help themselves. They remember what time of year these walnuts ripen."

These crows, according to Geir, have also found an ingenious way to open the walnuts' hard shell: they drop the nuts onto the asphalt, much to the despair of local car owners who have come to expect a few dents in their vehicles when the walnuts miss their mark.

Geir isn't someone to be content with speculation, so he and his colleagues set up an experiment to accurately test the crows' memory. Again, they used the crow's hunger for eggs. A test crow was released into an area where the previous year it had found, and plundered, the eggs from several nests. The first thing it did was check those very same places again. Obviously, it remembered the exact nest spots. Only after it had gone through all the places where it had previously found eggs did it start to investigate the rest of the area.[45] The crow's talent for remembering nesting places is one probable reason why so many small birds go to the trouble of building new nests every single year.

Good memory is one trait that crows share with primates. Another is the capacity for social living. Many species of both primates and crows live in groups. To be able to build and understand social relationships, it's important to be able to understand how others within the group think. The crow world is a social world, full of friends and enemies, harmless idiots and potential robbers. An individual crow's success depends on its ability to understand and act smart within this social environment.

Whatever the reason, evolution has given crows an extremely large brain in proportion to their body weight.[46,47] As large, in fact, as in our closest relatives, the chimpanzees.[48] The parts of the crow's forebrain, namely the nidopallium and mesopallium, thought to be analogous to the mammalian prefrontal cortex, are especially large.

Crows can actually be even smarter than their brain size indicates. As with a number of other songbirds and parrots, crows have smaller and more densely packed neurons in the brain than mammals do[49]—perhaps because they need to be able to fly and must therefore be as light as possible. So, the crows' small brain has room for a lot more neurons and interconnections than a similarly sized mammalian brain.

No doubt: crows are smart. And in the cities that means *street* smart. Urban environments are among the most unstable environments on Earth. Since the 1950s, between 70 and 80 percent of the green areas in Norway's cities have disappeared, and a similar pattern is seen in many cities around the world.[50] Buildings are demolished, they are constructed, the people living in them develop new habits and use new technology. I imagine the daily life of a crow involves a range of new problems like: How can I get under the lids of the municipal garbage bins? How can I make holes in the takeout boxes from the new Thai food place on the corner? How can I avoid getting run over by the new and eerily silent electric cars? Which of my human neighbors will feed me and which ones do I need to stay away from?

But what I wonder most is, how does it feel to be a crow?

When I experience something unpleasant, I become scared and angry. If I think about the situation later, I will feel the tension rising in my body. And if I meet the person who made me angry, my body will prickle with irritation whether I want it to or not.

Had the crows on the Seattle campus felt something resembling anger, or fear—like a child encountering the school bully in the corridor—when they saw the infamous Neanderthal walking beneath the trees? Do crows daydream about the summer heat in the middle of winter, when they gather in the trees, cold and hungry?

"They have feelings, of course," Geir said. Our crow trip had led us down to the beach. We had just watched three crows taunting a pug that was leashed outside a shop. I had noticed that the sounds they made as they tried to pull the helpless dog's tail were short and sharp, unlike the more drawn out *craaw* sounds they made with when the dog owner chased them away.

"Like many other animals, they feel pain, fear, and hunger. But emotions? That's something else." Geir sighed. "In the modern Western world, we see humanity as being separate from everything else. We see humanity as nature's shepherds; we are superior to it, separated from it. We find it hard to believe that creatures that don't resemble us can feel emotions the way we do. We might go as far as believing that chimpanzees can be happy and sad. But why can't crows be happy and sad?"

He smiled. "At the same time, it is easy to ascribe human emotions to other creatures. We don't know how birds are experiencing the world. I've yet to find a technique for interviewing a crow."

Geir's crow-research colleague, John Marzluff, hasn't interviewed any crows either. But he has let their brains speak directly, in a language that humans more easily understand—technology. In a follow-up experiment to the Neanderthal session, Marzluff and his colleagues went round catching wild crows while wearing caveman masks. The crows would therefore have perceived these masks as scary. During their captivity, the crows were fed by humans wearing other masks—which the crows were meant to perceive as kind. The researchers then used a PET scan to view the crows' brains while some of the birds were shown humans wearing the "scary" mask and others were shown humans wearing the "kind" mask. The idea was to see which of each crow's brain centers was active.

When the crows were shown the scary mask, the amygdala and cerebral cortex were active.[51] In mammals, these centers are associated with fear. However, when the crows were shown the kind mask, other parts of the brain became active. In mammals, similar brain centers are associated with motivation and hunger.

The researchers conclude that there is a strong similarity in how the crow brain and mammalian brain uses the areas that process things such as emotions, when we recognize, and emotionally deal with, a face. They believe it's possible that these systems in the bird and mammalian brains have the same origin.

In an interview with the Norwegian newspaper *Aftenposten*, Marzluff elaborated:

> The patterns we saw in the crow brain were exactly the same as those we would see had we scanned a human's reaction to a friend, an enemy, and a neutral person, respectively. Although it's three hundred million years since we shared ancestors, our brains still have some common traits in terms of structure and function. Both crow and human behavior are based on perception, emotional state, and experience. When we see an old friend, we get a warm feeling because of the chemical reactions in the brain and hormonal system, and when we see an enemy, we experience fear and anger. Crows feel the same things.[52]

Crows might have feelings. But it is beyond doubt that humans have feelings about crows. While working on this book, I have heard a huge number of crow stories. They are often along the lines of: "I don't particularly like crows, but..." I have listened to my hairstylist talk about his neighbors' tame crow so enthusiastically that he totally forgot about cutting my hair. I've heard about crows taking daily walks with old ladies, crows following hedgehogs across the road, crows vengefully eating the pet goldfish of people who suddenly stop feeding them, crows bullying cats, and crows using a snowy rooftop as ski slope for no other clear reason than because it's fun.

Whether Marzluff and his fellow crow researchers have found the absolute truth about the crow's strangely recognizable behavior I cannot say. As with another two-legged city-dwelling species, the crows' way of life has not been completely dissected into pieces of behavioral psychology and evolution.

Geir and I had by then walked through a small wood full of tall, straight-trunked pine trees and arrived at the sea. The snow almost reached down to the shoreline. A smell of rotting seaweed, along with the cold air, stung our nostrils. The sea was flat and calm: small waves lapped against the rocky beach and the giant blue-gray boulders. Geir then told me, almost tenderly, about a crow that had settled right beside a café where he had been sitting enjoying his coffee. It had found a piece of white bread but wasn't happy about the consistency.

"So, the crow dipped the bread in a nearby puddle, like it was dipping a flaky croissant into a cappuccino," Geir said, demonstrating how the crow repeatedly turned the bread while evaluating it with a critical eye, to see whether it was soft enough.

In the end, when the crow looked satisfied, it hopped up onto an available café chair to enjoy its food.

We stood right by the shore and watched a crow picking through the mounds of seaweed. Its beak was a solid, beautifully functional, and well-maintained tool powerful enough to cut through plastic nets and sensitive enough to pluck snails from their shells. Its wings and head were both dull and shiny, as I imagine black pearls perhaps look. Geir explained how it is just before mating season, as it was then, that the crow's plumage is at its best. A new and spotless crow uniform. It was the first time I had seen a crow and thought that it looked beautiful.

Geir was due to retire from his job at the university in two months' time, and was looking forward to what he described as "writing for eternity." He was, of course, going to be writing about crows.

The wind blew onshore from the south as we walked. It had been picking up throughout that short January day and, just after sunset, had developed into a steady breeze. The horizon was an ice-cold pink. Geir and I said goodbye. I watched as he strolled down the path, back to the articles waiting for him at home, while I thought about my strained relationship with crows. Maybe some of my aversion to them stemmed from the fact that they don't waste any time or effort on being liked. They are not set to appeal to others by having nice feathers or voices. They don't beg for human attention or food the way that mallards or street pigeons do. Crows ask for nothing. They just take what they want, as if it's their absolute right to do so.

That winter I started to carry binoculars everywhere. I could stand for a long time watching the crows in the huge trees downtown, and they often did things that made me want to call out to the nearest passerby: "Look!"

One late winter's day, I was waiting for a bus and witnessed two crows struggling to get a thin, foot-and-a-half-long twig up into a relatively dense medium-sized rowan tree. One of the crows had the unwieldy load in its beak, and after a struggle involving several failed attempts, it finally reached the trunk at the center of the tree. Using their beaks and claws, the crows then placed the twig in a tree fork, where it sat, rocking gently, while one of the crows guarded it. The other crow flew down to the ground and collected a new, slightly thinner twig. And with the new twig in its beak, it then fought its way back into the tree. The two crows looked like a couple of builders as they helped each other maneuver the second twig under the first one. You could almost imagine them giving crow commands to each other: "Right a little! Up a bit!"

It was a while before the crows finally managed to get the two twigs to sit crosswise.

Was this the start of a crow's nest? I had seen hundreds of crow's nests in my life without ever considering how they get built. I couldn't help wondering how the crows knew where to place the first branch, and then how to keep building to give the nest a secure foundation.

It must have been after seeing this construction that I began to look in amazement at the crow's nests I passed—just the fact that they were *there*, like huge black scribbles in the trees. As far as I could remember many of them had been there for years, and would have to have been skillfully built to endure the strong winds and weight of heavy spring snow.

I began to treat crows and ducks equally when I fed them bread by the river. It surprised me that, although they were about the same size, the mallards had the upper hand. If there was ever a tussle between a crow and a duck over a piece of bread, the crow would back away.

Still, I had no illusions about the crows. We were not friends. I knew they would peck my eyes out given the chance. But I still took binoculars out with me whenever I left the house.

The Night Singer

I T WAS 5 A.M. and a blackbird was singing as I stood inside the fence surrounding Oslo's botanical garden. The gates were locked at night, but I knew a place where you could easily jump the fence, and the gardens were even nicer when you weren't supposed to be there.

There had been a touch of spring in the air recently. The asphalt paths were slippery from the brown, wet leaves which had laid there since the previous fall. Beyond the cones of light from the streetlamps, I could just make out the contours of several huge trees. The sky was no longer pitch black but dark blue. Between the melodies of the blackbird's song, I heard a car passing by, the pounding bass from its speakers growing fainter and fainter as it vanished up the street.

I had been listening to the blackbird's song for a while already. It sounded almost like free jazz; flute-like notes followed by discordant sounds and wild tonal leaps. Melodies, followed by pauses, like an orator wanting to emphasize the point they had just been making.

The singing followed me as I continued through the garden, past leafless hedges and small ponds. But the bird was nowhere to be seen. Until I reached the fence on the other side of the garden. There, just beyond the fence, sat the jet-black bird. It looked at me through its round black eyes, while pointing its bright yellow beak upward at an oblique, defiant angle. It then hopped a couple of times across the brown grass and took off. Very few birds are active at night, which might explain why watching it fly into the twilight felt so unnatural.

The singing, however, continued. Listening to birdsong has never been a habit of mine. The song has just been there, like a soundtrack to the spring, until, at some point in early summer, it no longer is. But that night there was no rush, and nothing felt more important than anything else did. That's perhaps what compelled me to stop beneath one of the beech trees and wait for a few minutes, completely still, listening to the soft tones while the sky grew brighter overhead.

The blackbird's song followed me all the way home—along several blocks and deserted streets that somehow had a touch of the forest about them—until I reached my apartment building. And when I awoke the next day, the first thing I thought of was the blackbird. Even while making coffee, I continued to wonder about this bird that had sung in the dark. Why was it awake in the middle of the night? Who was it singing for?

Wanted: An Experienced Partner

"If you hear the ping of a text message come from a bush, it doesn't necessarily mean there's someone in there spying on you. It can also be the sound of a blackbird that has been inspired by an iPhone."

These words came from a man, possibly in his late forties, speaking with a calm Mid-Norwegian accent, and with a long black hair braid dangling from under his cap. About fifteen of us had turned up on this drizzly March morning, most of us wearing Gore-Tex trousers and sturdy footwear. Standing beside me was an elderly woman holding the largest telephoto lens I had ever seen in my life, and a couple in their sixties who were having a loud conversation.

After telling my friends about the blackbird experience I'd had in the botanical garden, many of them said they had encountered this bird, or heard its melancholy song, while walking home after a night shift or drinking binge. None of them could tell me why the blackbird had been awake at night. And in the discussions that followed, I realized how little I actually knew about *why* birds sing.

I had known people while I was at university who were fanatically interested in birds, so much so that they struggled to finish their degrees: all the final exams were held at the end of the spring semester—at a time when many rare birds can be seen and heard all over the country, usually in the most remote areas. So, while the rest of us completed our multiple-choice assignments about biological diversity, they were lying behind a bush somewhere, delighted at having an especially rare woodpecker or flycatcher in their sights.

I had always wondered what made birds so fascinating to these bird-watchers, and finally I had the necessary urge to investigate. And so, two weeks after my blackbird experience, I had boarded a train to Levanger, a small town northeast of Trondheim, to learn more.

ON THE FIRST DAY of the course about birds and birdsong, the air in the park was humid and the trees were shrouded in mist. I had a pair of binoculars hanging around my neck. Audun—the man with the braid, who happened to be leading the course—had already warranted a quick snoop on the internet, which revealed he had been a didgeridoo player at one time. I suspected the braid was once part of a more extensive hairdo.

The blackbird, Audun told us, is extremely good at imitating sounds and weaving them into its song. Blackbirds are known to have mimicked ambulance sirens, meowing kittens, creaking doors, and the song "Barbie Girl" by Aqua. But it isn't the only bird that's good at doing this. Many bird species weave the sounds they hear into their song, and the best imitators can hit the note and the timing perfectly. One local newspaper reported how a starling had been mimicking the guard's whistle at a local train station, causing the trains to leave prematurely.[1]

BIRDS DON'T ALWAYS DO what humans want them to do. Researchers trying to teach a group of starlings new melodies by playing the tunes from a tape recorder, found, to their disappointment, that the starlings refused to learn them. However, after a while the starlings' song included the sound of a cassette player being turned on, the hissing of audiotape, and the coughing of researchers.[2]

THERE ARE MANY EXPLANATIONS for why a lot of bird species imitate sounds, Audun said. But one might be that they are demonstrating what life experience they have. In a good song, a bird will combine the sounds it has heard in its life. The longer it has lived, the more new sounds it will have heard. Birds that winter in other parts of the world will often hear the songs of other bird species (not to mention other car alarms and ringtones) and then weave these exotic elements of the new sounds into their own song.

So, a song containing a lot of mimicry is the bird world's version of an Instagram account replete with travel-bragging photos. It's a way for the bird to show off all the places it has visited. Birdsong tells competitors and potential mates that the singer has had lots of time to learn new melodies, which indirectly means that it is good at surviving—an extremely sexy attribute in the bird world, and just the attribute birds would like to pass on to their offspring. In some species, birds that sing the most complicated songs find themselves rewarded by the opposite sex—with extra copulation.[3] And because of the variety of sounds in the city, humans have given birds an entirely new arsenal of songs to copy melodies from. For example, one of the great earworms of the '90s, the Nokia ringtone, apparently became popular among Copenhagen's starlings in the early 2000s.[4]

In his book *The Sound Approach to Birding,* author and birder Mark Constantine describes how life experience isn't the only thing that affects a bird's song; it is also the bird's mood, or more specifically how horny it is. Many adult songbirds have a non-territorial song that they use before mating season begins, which is typically quieter and not intended to trigger aggression in other birds. When a bird is looking for a mate, however, its song will be full of sexual energy (it saves the dynamite until it's needed).[5]

THE WOODED COPSE we were standing in was full of such information. Spring was lagging behind up here in the north and the trees were still leafless. The woods, however, sounded like a symphony orchestra, albeit without a conductor and with each musician relentlessly playing their part. Occasional movements above us revealed that the sounds weren't coming from the trees themselves. A branch quivered and the raindrops clinging to it fell to the ground. Something had landed, or flown away.

One of the people in the course, an elderly man, said "*shhhh!*" and pointed toward a tree from which pieces of melodies cascaded. They seemed familiar, but I still couldn't quite place them.

"A chaffinch," said Audun. A ready-to-mate chaffinch has a consistent and easily recognizable melody; it was one of the few bird calls I had memorized. I listened again and could actually hear bits of the chaffinch's characteristic descending trill, combined with fragments of other melodies. Audun explained that the young birds of many species sing differently to the adults. They babble away in a constant stream of sounds while trying out new melodies; they are more interested in testing their own vocal skills than listening for answers. The chaffinch we heard was perhaps one of last year's hatchlings and hadn't yet settled into the species' typical song.

Under the naked trees, a combination of rasping then delicate then soft tones formed a palette of *kuuu-it*'s, *twait*'s, and *zzzitt*'s. It felt like we'd been engulfed in a tight-knit exchange of communications, and yet we understood just a tiny fragment of them. Our disorderly group had fallen silent with studious looks on their faces. Even the chatty couple had gone quiet. Perhaps because we struggled to grasp the fact that each bird species—each individual bird, even—has not just one but a whole repertoire of songs, loaded with information about both the sender and the context.

I REMEMBERED FROM my biology studies that males do all the singing. It is their way of letting other males know that the territory is occupied—while also showing any females how virile they are, and that the females are welcome.

But it turns out that the story is more nuanced than that. Not much is known about female birdsong, mainly because so few people have been looking for it.[6] In species where the males and females look very similar, the singing individuals have often automatically been considered male.[7] If the bird is singing, it must therefore be a male. And so, only males sing.

Another reason that female birdsong has been so easily overlooked is because their song can be different from males', or because it sometimes only happens at certain times during the mating season. Whenever researchers *have* intercepted the female's song, the song has often been described as an anomaly caused by, for example, unusually high levels of sex hormones in the female.

Recent studies show that over 60 percent of songbird species may have singing females.[8] This finding suggests that many of the females play a more active part in the mating circus of the bird world than was previously thought. Like males, females sing for several reasons: female starlings might sing to attract a partner, for example, while female sparrows sing to defend their territory from other females. In a number of species, females sing duets with their mates. One hypothesis for this partner singing is that they want to show that both are defending the territory, to dissuade would-be intruders from challenging their right to it. Another possibility is that both the female and the male are singing out of jealousy: when one of them calls out enticingly, the other will respond to show other admirers that their potential partner is already taken. Duet singing is perhaps the bird world's equivalent of a wedding ring.

Young birds depend on listening as well. For example, anyone who has been near a busy tern colony will know how chaotic and tinnitus-inducing it can be. It is almost impossible to single out, let alone recognize, the sound of an individual tern. But the fledgling terns in the nests are able to recognize their parents' calls among the chaos of wings and beaks overhead—and will respond so that their parents can offer their precious food to the right beak.[9]

A Noisy Home Office

The blackbird I heard singing in the botanical garden was a jet-black male (the name blackbird is actually misleading: half of these birds—the females, that is—aren't black but brown). And its song was so varied it may well have been an older bird. But who was it actually singing to at such an hour?

Audun told us to close our eyes. The excited giggles coming from the group revealed how long it had been since any of us had played with our senses like this. I did as I was told, and noticed how the surrounding noises became more distinct. I heard the loud hiss from a passing truck releasing its clutch. A car with a broken exhaust pipe roared past, and a crow cawed raspily. A blackbird's song was cut short by the honking of a car horn. Levanger may be a small town, but the air was still full of human-made sounds even at such an early hour. The crow and the blackbird's songs merged with the urban noise, and became a part of it.

The voice of the city is the sound of thousands of engines, conversations, lawn mowers, construction workers, and seagulls, all coalescing into a never-ending drone. While the forest will muffle sound—because trees and leaves and vegetation absorb sound waves—hard surfaces will reflect it, making any noise last longer.

Noise can be harmful to humans. In urban areas, many people live with noise levels that can have a negative effect on their health.[10] This noise also affects birds. How can a blackbird know whether an intruder is an aggressive adult bird wanting to fight over territory, or a frightened young bird, if it cannot hear the way it is singing through the cacophony of human-made sounds? A misunderstood song can result in a fight where both parties risk being injured or killed unnecessarily.

For city birds, navigating the soundscape must be like having an important job where you're bombarded with messages and forced to sit in an office with a terrible internet connection. You send emails but have no idea if they've arrived, and you only receive fragments of emails in your own inbox.

How do city birds get round this problem?

The background noise of a city has a low frequency, which to humans sounds like a deep, steady hum.[11] The city's sopranos, such as the European robin, manage just fine—these birds sing at a higher frequency than the city noise and avoid being drowned out as a result. Others simply turn up the volume. According to a German study, Berlin's nightingales have become louder in response to the noise from traffic.[12] On weekdays they sing at a higher volume than on weekends when the roads are quieter.

Birds that sing in lower tones, and are unable to turn up the volume, have a bigger problem. What's the point of singing a deep and moody love song if no one can hear it?

Some bird species solve this problem by moving the song up the tone scale. The researcher Hans Slabbekoorn compared typical city birds with their country cousins and found that city birds sang with a higher frequency in ten out of ten(!) cases.[13] By shifting to a higher pitch, city birds make their song easier to distinguish from street noise, in the same way that a piccolo can be heard clearly over the tubas in an orchestra.

Various studies have supported Slabbekoorn's observations. The species that thrive in the city are often those capable of adjusting their vocal range to a frequency that doesn't get drowned out amid the noise.[14] And it's not only birds that adapt to city noise. The research suggests that urban squirrels, too, may have adapted in order to communicate. While much non-urban squirrel communication is done through sound, city squirrels depend more on visual signals, such as wagging their tails when, for example, they want to alert others.[15]

Some studies indicate that city birds adapt to noise when they encounter it; they sing more brightly when the noise starts. Other studies suggest that there may be genetic reasons why some bird species sing more brightly in the cities.[16] The noise of the city has, so to speak, been etched into the city birds' DNA.[17]

AS WE STOOD among the trees in Levanger, the riot of birdsong was audible well above the traffic noise. Audun was able to pick out species after species, like a magician identifying cards in a sealed deck. He had trained his entire life to recognize the different bird voices. It amazed me how the ear can be so finely attuned that it can distinguish between nuances of sound. But it also gave me hope that even I might be able to understand more of the information birdsong contains. And it's perhaps no more strange than the fact that humans can easily tell hundreds of human voices apart.

Audun showed us how the tones and overtones of birdsong can be written down in a form of music notation. These so-called spectrograms help us actually "see" birdsong. The horizontal axis represents time, while the vertical axis shows the pitch or frequency. Wavy lines and shapes indicate a melody with a rising and sinking pitch. And the broader the line, the louder the volume. Any overtones that modify the timbre of the birdsong are represented as static running parallel to the melody lines. In books and on websites about birdsong you can often see the sounds in addition to hearing them. This visual notation allows us to involve an extra sense to help understand the song's structure, Audun explained.

IT THEN STARTED pouring rain, and the birdsong dwindled. When Audun eventually signaled that we could retreat to our cars, I sighed with relief.

A FEW DAYS LATER the course was over, but for me it was just the beginning. Friends and acquaintances no doubt saw then how single-minded I'd already become. After returning from Levanger, my attention was almost constantly directed at the trees. I spent my spare time listening to pre-recorded birdsong, memorizing it, and testing myself. It was like the deep immersion you experience when learning a new language, one you really want to understand, and you start to feel like you're on the cusp of unlocking its secrets.

Almost every day I sat under the beech tree in the botanical garden, with a pair of binoculars and something to draw on, while surrounded by a riot of noise. The noise was especially loud when the air was humid following a rainy spell. And I found that the longer I sat there, the more birds I could hear—either as the birds got used to me being there or as I became more tuned in to them. Or both.

After having sat for a few mornings, I began to notice a difference in the timbre of bird calls. While the great tit's song had a metallic ring to it, the coal tit's song sounded more like a plastic whistle. Blue tits, the little hotheads, have a faster, sharper sound. The black-and-white fly-catcher's song is structured similarly to the great tit's, but the sound is completely different; more like a washcloth being rubbed vigorously on a windowpane.

I also started to draw my own somewhat rough-and-ready spectrograms. Having a visual record of a birdsong's structure was very useful, and my notepad gradually filled up with scrawled attempts at capturing the logic and systems behind the sounds made by blue tits, cuckoos, and wrens.

With a couple of bird calls jotted down, I had a reference work to compare others with. More breathless than the great tit? Marsh tit. Like a nervous and shaky blackbird? Mistle thrush. Like a clarinet choir on acid? A flock of whooper swans.

One of the weirdest sounds I heard came from a bush. It was a mix of affectionate clucking and babbling sounds unlike anything I had heard before. I moved closer and there, under the bush, sat a magpie looking up at me innocently. For the first time, a magpie had revealed its affiliation with songbirds to me. And there I was thinking that I at least knew magpies.

Unofficial spectrogram of a blue tit

THROUGHOUT MAY, the activity in the trees above my seat in the botanical garden increased. With my attention drawn upward, I saw and heard several bird species I had never seen before as they returned from their winter retreats in the south. I had no idea there were so many birds in the middle of town. I also began to notice other bird-watchers wandering around in the botanical garden too. It was like we were all part of some mysterious secret club whose membership was revealed by having binoculars in your inside pocket and being fixated on the trees. Our silent fellowship reminded me of Pokémon GO players who walk around staring at their cell phones, detecting nearby characters that ordinary people can't see.

The sound of a reed bunting,
loosely interpreted by the author

As a child, I had a powerful urge to learn about animal-tracking and signs. I dreamed about becoming like the Native Americans I read about, who could read the natural signs around them and understand things that cannot be seen directly. The trouble was I lived in the city, so in the forest I was only ever a visitor. I had never been there long enough to notice the difference between dog tracks and fox tracks, or between the feathers of an owl and those of a fieldfare.

I had now found another entry point. It felt as if I had gained access to a snippet of the information network stretching across the parks and streets. Recognizing a little bit of birdsong made me feel almost euphoric. Perhaps it's similar for graffiti artists who can understand all the tags spray-painted on a city's walls, or for architects who can walk around a city and read its history in buildings and parks. Either way, for me it was birds.

AS I SAT AND LISTENED, heavily focused on the tones and overtones coming from the foliage above, I began to understand something Mark Constantine mentions in the closing chapter of *The Sound Approach to Birding.* He says that the most important thing perhaps isn't that you're able to distinguish between species, between females and males, between horny teenagers and well-traveled seniors. What you really learn from birds is how to listen.

The sound of a crow, as perceived by the author

I still hadn't got an answer to my question about the blackbird that was singing in the middle of that dark March night.

WELL, THE FIRST THING I found out as I sat with my binoculars under the beech tree in the botanical garden was that there were blackbirds everywhere. Every sixty feet or so a black male or a brown female was pulling up a worm or plucking old rowanberries from the lawn.[18] And they spent a considerable amount of time sitting in the trees singing their hearts out, often in duets. What I had initially thought was a lone blackbird pursuing me that night in the botanical garden may well have been three or four blackbirds, each one sitting in its own territory and singing, listening to the other birds, and singing some more.

The other thing I learned was that the city has rhythm. I could sit under the beech tree and hear how the character of the traffic noise changed as the morning rush hour subsided, and how the sound of people in the park increased as lunchtime approached. And when I refer to the "city," I don't just mean the traffic and the people but the city's birds as well. After a downpour of rain, for example, the sound of birdsong was intense.[19] The singing would reach a crescendo in the morning and then take a break for a few hours in the middle of the day before starting to increase again just before sunset.

ONE NIGHT I set my alarm for half past two in the morning. When it rang, I shook myself awake, put my clothes on, and staggered out the door. It wasn't a totally dark night: the glare of the streetlights gave the sky a dark-gray cast. Parked cars lined the empty, dimly lit streets. And the wind whistled in the trees as the faint throb of bass notes blew in from a faraway party. I had almost forgotten that the city could be so quiet.

Before I even got to the park, I heard a blackbird singing from a rooftop.

Sitting under the beech tree this time felt less unusual than on my first night in the park. Soft melodies wafted under the branches from either side as the half-light grad-ually shifted to blue. It was on this night that I found out why the blackbird has gone from early bird to night owl. As I sat under the beech tree, it felt like I was eavesdrop-ping on the blackbird's private chat room: the night. Not only do city blackbirds switch to a higher frequency in order to talk to each other in peace, they have also found their own rhythm: they start singing up to five hours ear-lier than their cousins in the forest.[20,21]

The War on Ants

About Sex, Cohabiting, and Messy Windshields

IT IS DURING WARTIME that our values are tested. Our war began suddenly on a Tuesday night in April, after the frost in the flower beds had thawed. There were fewer dirty snowpiles in the streets. And the evening sun, as it shone through the unwashed windowpanes, cast mottled patterns on the kitchen floor. Someone shouted in the backyard and hoarse laughter rose above a rooftop, giving voice to the restless excitement that early April brings: confirmation that spring is underway and that you're already lagging behind. I always find this sense of anticipation makes me feel uneasy: I want to go outside and be a part of spring's deluge of happenings, but I never quite know how to.

This particular evening was full of that same uneasy anticipation. I was in the kitchen buttering some toast for supper while waiting for something to happen. And as I took the bread out of the drawer, something odd caught my eye. The usual white crumbs in the bread bag had black crumbs among them. Crumbs that moved.

When I took a closer look, I could see how many there were. Tiny little dots, all over the bread, inside the bread bag, and on the kitchen counter. Ants! I watched as one of them raced toward the kitchen cupboard. It was a systematic attack. The invaders had built an entire road network of streaming black dots that stretched all the way along the shelf containing baking ingredients, took a detour via the oatmeal, then forked toward the sugar bag and the flour bag. White sugar grains danced just above the floor, along roads that branched off from one central highway that vanished through a gap under the front door.

We'd had insect invasions before, of course, from fruit flies and flour moths. Every fall, we are visited by a large house spider that lives behind the couch and scares the crap out of us, until we just get used to it being there, at which point it usually just disappears again. But this invasion was bigger, and it felt exciting in the same way that a storm or a power outage, or any event that interrupts normal life, feels exciting. So, I studied the stream of ants while pondering how easily the line between the outside and inside world had been erased.

I'm not sure if I like ants. As a child, they frightened me. Under the loupe, which I borrowed from my grandma's dusty drawer, they always got too close. They would raise their upper bodies from the ground and fence at me with their antennae. I was used to most animals being timid, and to the idea that having a connection with animals was about building trust. With ants it was the opposite. They would attack whenever the opportunity presented itself.

I stood in the kitchen wondering if that was why I still had an aversion to ants: they are courageous to the very end. There's nothing I can do to frighten them.

Even when it comes to humankind's favorite pastime—world domination—we can't claim any clear victory over ants. For every human on Earth, there are over a million ants.[1] If success is calculated by total biomass—well, some believe that the combined weight of all the world's ants is greater than the combined weight of every human on Earth.[2] I imagined my body slowly dissolving into a mass of swarming black dots. Was there a swarm of ants equal to my own body weight at the end of this ant trail?

Ants are found in huge numbers, in most habitats, on every continent except for Antarctica. Almost anywhere you are, you can rely on ants being there.[3] That's the other thing I can remember from my expeditions with Grandma's loupe. There wasn't much else to look at. Frogs were difficult to catch. So were flies. But ants were always there, predictably furious.

Back in the kitchen, the ant trail was now streaming out onto the veranda. And so I walked down the back steps and opened the door to the garden to see where the ant trail led. And in the dwindling light of the spring evening, I saw a procession of ants winding its way along the uneven stone wall beside the veranda from the second floor down to the ground.

Beneath the veranda, the ant trail dissolved into the soil. After sundown the heat from the southwest-facing wall keeps the flower bed warm, and in the spring it is the first patch of earth to thaw. A week earlier, I had planted a few spindly bean plants there which I'd tied to a wooden trellis. That's where all the ants were disappearing. I bent down to find out where they were going and saw that many of them were darting into a hole beneath one of the slabs.

As I stood there, I noticed something strange about my beans. These plants, which had been bursting with sap a few days earlier, were now hanging limply from the twine. The leaves had curled up at the ends, and the angular stalks were covered with tiny black buds that became increasingly dense toward the top. These buds, which looked like teardrop-shaped blisters, were tapered at one end with a powdery black texture similar to a dark plum. They were black aphids and they protruded from the bean stalks with their rear ends aloft. I grimaced and crouched down for a closer look. The top of the plant stem was so infested that when one of the aphids drowsily changed position, all the other aphids' soft bodies rubbed against each other.

Fascinated and repulsed, I looked down the stem and saw a black ant on its way up, running at full tilt while feeling with its antennae as though it was blind. More ants came running up the stem behind it, although running probably isn't the right word. Running presupposes an ability to slow down to walking speed, and from what I could see these ants had just the one gear: full speed ahead.

The ants stopped beside a cluster of aphids (again, stopped isn't the right word because the ants' forelegs and antennae were constantly moving) and stroked those aphids with their antennae. The whole stalk was a mass of antennae and legs. The ants were more than twice the size of the plump little aphids, and way faster. As the ants swept their antennae over the aphids' scaly bodies, their automated movements looked like factory workers on an assembly line.

But what happened next was the strangest thing I had seen in a long time. A tiny bubble emerged from an aphid's rear end. It contained a clear liquid, and it expanded, until after a few seconds the shiny droplet was about a sixth of the size of the aphid itself. One of the nearby ants then stuck its pointy head into the bubble, which became smaller and smaller until finally it was gone. Had this ant been drinking from the aphid?

The sun had set behind the neighboring apartment building, and since I was only wearing a T-shirt I began to get goose bumps on my arms. The flickering light from a TV streamed from a first-floor window, and two young girls stood on a veranda talking excitedly. The scent of dog poop wafted over from the snow piled up near the bike racks. Everything seemed quite normal. But that spring evening had produced some quite unexpected events.

When I returned to my apartment, I started googling "how to kill ants." But I also began to wonder: How did all these ants find their way into my kitchen so quickly? And what on earth was it that came out of the aphid's butt?

There's an old, possibly Chinese, saying that goes: "Know your enemy." But I knew very little about mine.

Ants have simply always been there. They are the annoying things that get in your hair when you're lying on the grass in the park, that you quickly brush off your clothes, and that live under the slabs in the backyard. At primary school, the kids who liked ants were often the same ones who kept pet spiders and stick insects, which they loved to put in the hair of unsuspecting classmates. The ant lovers in my class were annoying themselves—they were not potential allies in schoolyard battles.

But it was more than springtime curiosity that stopped me from letting the ants have it. I had previously criticized family and friends for using insecticides in their gardens. It would be hard to admit to spreading poison around my own backyard and hear everyone thinking: *Oh, our self-righteous idealist has had a reality check.*

And so it was with some (self-)satisfaction that I decided to make peace, temporarily, with the ants. It would be a while before anyone with a low tolerance for kitchen irregularities (i.e., my mother) might be visiting. And besides, the ant invasion was exciting; it was something new and in step with the approach of spring. As long as the ants didn't cross any more boundaries, I would tolerate them being there.

Royal Privileges

Our truce was fragile, and it didn't last very long. But it was the ants that let down their end of the bargain first. One night, hundreds of them found out where I kept my Portuguese honey, a heavenly and very expensive honeycomb topped with a layer of tiny, cream-colored beeswax bubbles. The next morning, the whole jar was covered with a sticky layer of dead ants.

In the morning the ants were gone, but that night the black trails were back. As one of them scurried around on the chopping board, I caught it between my thumb and index finger. It squirmed and kicked, and I just managed to resist the urge to throw it away. Close up its little ant body looked dark brown, like a piece of highly polished ebony, and it was divided into three sections: a head, a midsection to which all its legs were attached, and a large pointy rear section. The midsection was muscular, like a weightlifter's torso, its legs and antennae were a lighter color and constantly moving, and its head was like a heart-shaped motorbike helmet. Its mandibles opened wide, eager to bite onto something. Despite my reluctance to do so, I crushed the ant between my fingers.

From its size, its brown-black color, and the fact that the squashed ant didn't smell citrusy (as jet ants do, according to AntWiki), I deduced that the intruder in my home had to be a *Lasius niger*, or black garden ant. I grabbed my loupe and took a closer look. And I jumped. The mandibles on this motionless ant corpse looked terrifying. Under the loupe I could also see the tiny hairs that cover this ant's body and give the genus its name—*Lasius* comes from Greek and means "hairy." I was dealing with black garden ants. And in the extensive literature I had managed to find on ants, including the book *Journey to the Ants* by ant researchers Bert Hölldobler and his partner E. O. Wilson, I found out that they were all female.[4]

Ants have a sex determination system that challenges my view of what gender can be. With many animals, biological sex is determined primarily by sex chromosomes (there are modulations that can change this expression, but that's another matter).[5] With ants, sex is determined in a different way. Ant queens produce roughly three types of offspring. Ant eggs that are fertilized by semen from the queen's sperm bank inherit a double set of genes: one from the queen's eggs and one from the male's semen. They become diploid, like humans. Most of these ants develop into female worker ants. Thus, the vast majority of the colony are diploid female workers, who tend the larvae, collect food, and fight other ants.

However, a queen can decide which eggs will be fertilized or not. She does this by contracting or opening ducts on her abdomen where she has stored the sperm of the males she mated with during the only sexual encounter of her life. So, when the colony has enough diploid female workers, and can afford a few males, the queen will close the duct to the semen. Any subsequent eggs will not be fertilized by the male's semen and will produce haploid males, with just one set of genes from the queen. These males are useless for anything other than sex.

The third category is the ants that will become young queens. I'll come back to these transformative figures in a bit.

A queen lays all the eggs in the colony and passes on all her genes to her offspring. Her female worker ants do not have their own young, and so only a part of their genes are spread, indirectly, when their brothers and queen sisters fly away, mate, and build new colonies.

Why don't the female worker ants rebel? And how does the queen manage to avoid coup attempts from rival females wanting to spread their own genes? Different ant species have found different solutions to this question. In some species, the type and the amount of food female ants receive determines whether larvae become workers or queens. In other species, the queen produces substances that stop female ants from maturing sexually.[6] The result in each case is the same: when a colony is established, the queen stops all the female worker ants from progressing past an early stage of development; they will lack wings and have undeveloped ovaries so they cannot threaten the queen's exclusive right of reproduction.

Later, when the colony is well established, the queen will relax her control over some of the fertilized eggs and allow them to become queens. When the future queen is fully developed, she has two options that worker ants don't: she can fly away and she can mate. Freedom and sex are the queen's royal privileges, but with privileges comes responsibility. Unlike the workers, who live simple lives performing routine tasks along with their sisters, the new queen in many species has to fly away, out among the dangers of the world, to give birth—alone—and start her own colony. The queen of a colony will not tolerate competition at home, which makes even the most domineering human matriarchs look good compared to an ant mother. The ants in my kitchen are all-sacrificing courtiers who do everything for their mother and queen.

The male ant's only mission in life is to be a "sperm-bearing missile"[7]—to hatch, fly off to find itself a virgin queen, fire off its sperm if it manages to get lucky, and die shortly afterward. If males do succeed in mating, their sperm has a long life. Unlike human sperm cells, which are quite helpless outside the body, ant semen can survive for years inside a queen ant. When it comes to efficient utilization of sperm, the ant queen is in a league of her own. She mates just once, yet can live and lay fertilized eggs for many years. One black ant queen is reported to have lived in captivity for thirty years.

Thousands of newly hatched virgin queens take flight from the ant colony every spring, hoping to find, and be fertilized by, recently hatched males from other colonies who are on the same sex holiday. After fertilization, the flying queens spread like dandelion seeds across large areas and, just like dandelion seeds, very few manage to find a good place to settle. Most of them die, either splattered across car windshields or eaten by birds. But due to their large numbers, some of them will be lucky enough to land somewhere suitable for establishing a new colony.[8] If she finds a suitable place, a newly mated queen will bite her wings off and burrow into the ground to start laying eggs, thus condemning herself to a perpetual existence as an underground matron.

Exterminating an ant colony can be a difficult task, it dawned on me, given the situation in my own kitchen. An abandoned nest under a stone slab in an old apartment building would be an amazing find for a young queen house-hunting for herself and her two hundred million sperm. Even if I did destroy the existing colony, new ant queens could come and take over the vacant territory next spring, and I'd be back to square one.

The Soda Trade

The black garden ant is one of the most common ant species in the towns and cities across much of the temperate
world. Typically, for species that thrive in the city, it is
a generalist—flexible when choosing a place to live—and
has spread from its native habitat in Eurasia to North
America, Africa, even to Hawaii. The black garden ant can
build colonies in places as diverse as rotting tree trunks
or under rocks, in damp river valleys, in deserts, or paved
backyards. And it isn't picky about food, which helps if
you need to conquer the world. This ant is not only partial
to the contents of kitchen cabinets, it will also eat other
insects, flower nectar, and the fatty seed appendages of
some flower species, which it in turn helps to spread. The
black garden ant is also among the world's oldest farmers.[9]

Ants capitalize on the fact that aphids struggle with
something that is rare for the animal kingdom: they have
too many calories in their diet. I can only think of one other
species that has that same problem.

NEW APHIDS ARE born like this: during the spring months
a female aphid might land on a bean stalk. She will then
push her long proboscis into the plant, until it reaches
the plant's vascular system, which is bursting with
carbohydrate-rich—but low-protein—sap. Once connected to the vascular system, the aphid will start giving
birth to live daughters, around five a day, that will also
connect to the plant's sap network. After a week's adoles-
cence, the daughters will start cloning themselves, at the
same rate as their mother, until all the available space on

the plant stem is covered in sap-sucking clones of the original aphid. It wasn't a surprise that my beans were wilting.

But aphids need proteins and other nutrients, in addition to sugar, and these are found in sap but only in small amounts. So, for aphids to get enough of these substances, they have to drink a lot of sap. You could compare this situation to a human trying to survive on nothing but soda. A moderately active person weighing 155 pounds would have to drink nearly 15 gallons of soda a day to get enough protein.[10] Aside from the countless toilet visits, this person would eventually suffer all sorts of other serious nutritional deficiencies. Not to mention a hefty sugar overload: all that soda would amount to well over 20,000 kilocalories per day. To get enough protein, this soda-dependent person would need about ten times more calories than the average human requires. And these extra calories would be retained in the form of body fat.

But aphids do not gain weight. They instead discharge the sap's extra calories through their anus, as drops of sugary water. And this sweet excrement, poetically named honeydew, is where ants come into the picture. Aphids can defend themselves against enemies by kicking with their hind legs, although compared to ants they aren't particularly fearsome. So, many aphid species use their honeydew as a currency, to buy protection services from the more battle-ready ants who will chase enemies away from "their" aphids. The aphid species that really adapted to this coexistence have outsourced nearly all of their defense measures to ants.

When one of my black garden ants meets one of the colony's aphids down there in the backyard, it will start milking the aphid with a series of gentle strokes from its antennae and forelegs. And the aphid will respond by secreting a drop of honeydew from its anus. Some aphid species have developed a dainty little wreath of hair around the anus to make the harvesting easier for the ant: the hair-wreath holds the secreted drops in place until an ant comes along and siphons it away. In one summer, a colony of black ants can harvest as much as four cups of honeydew from its aphids.[11]

We know these facts because of the ant researchers who have dedicated their lives to ants, one of the most famous being the biologist E. O. Wilson. A working day for a curious ant researcher like Wilson can look like this: one day in New Guinea, while waiting for a lift by the side of the road, Wilson found a colony of scale insects. These sap-sucking insects are closely related to aphids, and are often milked by ants as well. Wilson plucked a hair from his head and tried to milk the scale insect, which may have thought the researcher was a giant ant because it secreted a drop of honeydew from its anus. Wilson confirmed that the drop tasted sweet.[12]

E. O. Wilson wasn't the first person to taste scale-insect poop. The story of the manna, which according to the Old Testament the Israelites received from above after fleeing Egypt, may well have originated from the sweet excrement produced by the scale insect *Trabutina mannipara*.[13]

Removing honeydew is important for aphids. If honeydew is released onto the plant stem, the sugar will attract harmful fungi. So, by removing the sugar the ants are tending to the aphid's personal hygiene. When no ants are present to do the milking, an aphid will drop its honeydew onto the plant or onto the ground. City dwellers will often notice that if they've parked their car or bike under a tree, everything will be covered by a sticky substance.

The price aphids pay for this ant protection is that they have to sacrifice some of their freedom. Ants are fairly unsentimental about their livestock. For example, they will bite the wings off "their" aphids should they begin to show signs of wanting to leave. And if the ants run out of other sources of protein, they'll just slaughter a couple of livestock.[14]

TWO DAYS AFTER the ant invasion began, I returned to the backyard and picked up one of the flaccid bean stalks to look at the black bodies. Maybe this should be the day that I learn to milk aphids? I thought. I plucked a hair from my head and raised the bean stalk to eye level. With the tip of the hair, I tried to stroke the nearest aphid on its back and rear end while thinking about what I might say if a neighbor walked by. Hitting the target was difficult, so it was nearly all prodding. And the aphid, which at first didn't move at all, eventually just backed slowly away.

After several attempts, I had to conclude that I wasn't as good an ant imitator as E. O. Wilson, the ant researcher. However, working on my milking technique had given me ample opportunity to look at the ants and aphids up close. The mechanical way the ants moved around the aphids, how they greeted each other, soberly and repetitively, by feeling each other's faces with their antennae, all seemed quite ritualistic. What were they actually doing?

Ant Language

The ants always came in the evening, just before it got dark. And on each of the subsequent evenings I sat for a few minutes at the kitchen table watching the ants on the breadboard. I had a feeling there was a rhythm, a system, to what at first glance seemed like a disorderly chaos of streaming bodies, legs, and antennae. The ants greeted each other like blind people, touching each other briefly on the face or body.

Most species of ants have mediocre sight at best, and many species are almost blind. They, along with the majority of living creatures, navigate using molecules not light.[15] Their world, down in the nest in the extensive network of underground tunnels, is a world of chemistry. While the city for birds is a landscape of song and color, black ants perceive their environment primarily through molecules.

The ant's olfactory system (its sense of smell) is located in its long, sensitive antennae, which it uses continually to survey its surroundings, much like a blind person using a white cane. There are four hundred different smell receptors in the olfactory system, four to five times more than most other insects have.[16] Humans have just as many smell receptors, which means in principle we can smell a trillion (one million million) different odors.[17]

I couldn't help but wonder how the world appears through the ant's sensitive antennae. I imagined wearing magic gloves with fingerpads loaded with taste cells. The flavor and smell of every object I touched would be drawn in through my fingertips. With these gloves on my hands, and a bag over my head to shut out the light, I might just come close to experiencing the world of ants.

But how do ants tell each other about something like a cupboard full of food on the second floor of an apartment building? This well-developed sense of smell also helps ants communicate with each other via a system of pheromones, external hormones that many insects (and other species) use to communicate with each other. The response to a pheromone is typically "hardwired"; it is genetically determined.

Ant pheromones are secreted through two glands located at the rectal opening. Most ant species have between ten and twenty chemical "words," which can be combined to express many different messages. Ants are like an advanced mixer tap for chemical signals.

Many people have speculated whether humans can be affected by pheromones too. And I quickly discovered that googling "pheromones" will trigger a deluge of targeted banner ads for human pheromone perfumes and deodorants, even though I was primarily interested in ants. One little squirt, according to the ads suddenly plastered all over the websites I browsed, makes you irresistible. Many of us do actually have an organ in our nose that responds to some pheromones called the vomeronasal (Jacobson's) organ; it is an evolutionary remnant from a time when we used our sense of smell more than we do today.[18] However, whether this sense really is a shortcut to humankind's instincts and urges is disputed, to put it mildly.

Ants can further modify the meaning of their chemical message through body contact, making the number of possible messages they can exchange far greater than the ten to twenty purely chemical signals.[19] Perhaps what I'd perceived as frenetic touching between the ants as they met on the kitchen counter was in fact an exchange of these very messages. For example, a hungry ant might extend a foreleg and touch its sister on the ant equivalent of her tongue. This touch will stimulate the ant to regurgitate some food for her hungry sibling. The ant researchers E. O. Wilson and Bert Hölldobler also attempted this technique while poking around with their strands of hair. Their reward? A tiny portion of delicious ant vomit.[20]

The problem with simple signals like these is that they can be manipulated. There is no room for nuance or interpretation. If ants were people, I think I know what kind of people they would be: cult members. They unquestioningly follow orders, regardless of whether the context makes sense or not.

It's fun to fool someone who is unwavering in their belief about something. Ants are unwavering. And ant researchers like to fool them. For example, if an ant dies inside the colony, it won't make any fuss. It will simply topple to one side with its legs curled up. As long as the ant smells the way it should, its siblings will leave it alone. But as the body begins to rot, its smell will change too. And this new smell will tell the siblings to carry the body out to the trash (yes, most species live in colonies with their own garbage dump).[21]

Wilson and his ant-researcher colleagues studied the substances an ant releases when it dies, and began to apply this odor to *living* ants. When these poor living ants then tried to re-enter the colony, they were resolutely carried out by their siblings and thrown onto the colony's trash heap. The unwanted ants would kick and squirm and be very much alive, but none of those actions prevented their siblings from throwing them on the heap. The ant world is no place for discussion.

Anyway to collaborate it's important to be able to communicate. Genetic studies show that when ants began to live socially the number of smell receptors in their antennae increased dramatically.[22] The use of pheromones in communication is likely to have been an important reason why ants spread and became so successful on the planet.

IT'S DURING WARTIME that our values are tested, and mine didn't last very long. Yes, I was fascinated by the ants and their strange way of life, but I was mostly repulsed by them. The affinity I had gained for crows and songbirds just didn't extend to ants. Ants were somehow too mechanical. It was impossible to identify with them, and my aversion to them surpassed my desire to see myself as a passionate naturalist. Plus my partner threatened to move out if we didn't get rid of the ants.

On the internet there were legions of pest fighters ready to help. You could buy insecticides with names like Killtox and Maxforce. Even the rhetoric was warlike: "If you see a large number of ants, it could be a sign of an attack." We were under attack.

"You have to exterminate both the queen and the entire ant colony. The mound will exterminate itself from within." I was fired up. *Yes! This'll show 'em!* On my way home from work a few days after the invasion, I cycled past the hardware store; picked up a blue, yellow, and white spray bottle the size of a bathroom cleaner; and read the information on the back which said: "Long-acting oil permeate. Can be used on absolutely any type of insect." It also said: "Highly toxic to aquatic life with long-lasting effects." So, the spray bottle contained a substance toxic to ABSOLUTELY ALL types of insects, including bumblebees, honeybees, butterflies, the spider who lives under the sofa, as well as the beetles that live between the stone slabs in the backyard. All generously included on the product's list of potential victims (even though spiders aren't insects).

As I stood in the kitchen watching the ants streaming over the threshold, I was overwhelmed by a sense of power. Their lives were now in my hands. It was the same feeling you get when you pour Drano into a clogged sink and it smokes and gurgles as the chemicals do their work. But before fixing the ant problem once and for all, I needed to check something with an expert.

Dynamite and Life After Death

"By the time I was your age I'd already blown up a lot of trees." The entomologist and professor looked harmless enough in her boots and yellow rain pants, and with curly hair sticking out from under her hat. But appearances can be deceiving. Anne Sverdrup-Thygeson knows how to handle dynamite, and she is literally exploding with knowledge. It had been a few days since the great ant invasion and the spray bottle stood undetonated on the kitchen counter. Anne and I were in Svartdal, a wooded valley park in the middle of Oslo, walking the trail beside the river that was in spring flood. And we were going at an astonishing pace too. I was lagging behind Anne like a flailing scarf. I would later hear that she is a competent ultrarunner.

After an extremely wet start to the morning, the spring sun was just beginning to peek over the dark clouds. Large drops of water dripped from the branches onto the trail. Clusters of pale-green fiddleheads stuck up here and there: the ostrich ferns were on their way. The smell of decaying wood rose from the soil. We had to constantly step over the dead willow and aspen trunks that lay pell-mell along the river's steep slopes; the valley felt dense and moist like a jungle. Svartdal is one of the few places in central Oslo where dead wood is allowed to lie and rot without anyone coming to remove it.

We had talked about the complex life of trees after death, and Anne had elaborated on her relationship with dynamite. In one of her projects, she and her fellow entomologists had blown up several trees and allowed the shattered trunks to die. The reason for this tree massacre was to see which insects would move in to break down the dead wood. It turned out to be quite a few.

As we stepped over a rotting willow trunk, Anne stopped mid-sentence and bent down for a closer look. Soon she howled in excitement. I don't know anyone who gets as happy as Anne does about dead wood. She has written books and given lectures all over the world about the fascinating lives of insects, and how much we need them if the world as we know it is to continue. Dead wood—the professor told me while examining the trunk—is full of life, insect larvae, beetles, and fungi. More alive than the living tree ever was. She then picked off a loose piece of bark from halfway up the rotten log and cheered when she saw dozens of holes bored into the soft wood underneath.

"Could be a longhorn beetle!" she cried, with such enthusiasm that two ladies, walking their poodles at marching pace, became interested in what we were doing.

I HAD TOLD ANNE about my ant invasion and the spray bottle that was standing on the kitchen counter at home. But before commencing with the extermination, I wanted someone to explain to me what was happening with insects and our car windshields. When I was a child and my family went on road trips, I always kept a close eye on the windshield, examining every insect as it struck the glass with a delightfully revolting splat. The best impacts— the ones that delivered the biggest yuck-factor—were the bloody ones that left little red splatters behind. When I'm out driving these days, however, I can wait much longer before I have to wash the windshield.

As we stood over the rotting tree trunk in Svart-dal, Anne explained that the world's windshields are increasingly clean because large and important insect and pollinator groups are becoming extinct. "A range of individual studies have pointed to a clear and somewhat dramatic decline of insect populations in different habitats and regions."

In one study, researchers had been catching insects in Germany's nature reserves, and their findings attracted worldwide attention: three-quarters of the insect biomass—

the total weight of all the flying insects that were caught—had disappeared from nature reserves within the last thirty years.[23] It is comparable to the world being hit by a pandemic that wipes out the entire population of the United States, except for the people living in California, Texas, and Florida.

Humans need insects. They do many vital jobs for us, the most well known probably being pollinating crops and helping decompose organic matter like wood.

Perhaps insects need rebranding, since they're not best known as a resource. In many myths, the gods punish humans with swarms of flies, mosquitoes, locusts, or some other disease-spreading, harvest-eating, and generally troublesome species. In the Old Testament, at least four of the ten plagues inflicted by God upon the Egyptians are related to insects.[24] Here in Norway, in old Norse mythology, the demigod Loki turned himself into a horsefly in order to torment people.

In the Middle Ages, people sought both divine and legal help to defend themselves against plagues of insects: Pope Stephen VI wanted to drive locusts from the fields using holy water, while others tried to *expel* insects from churches or other properties using the powers of the law. The legal persecution of troublesome insects continued until the end of the eighteenth century.[25]

In recent centuries, the war on insects has escalated. Invasions of ants, locusts, and other wretched little bugs are no longer a divine punishment that humans have to endure. The twentieth century has been an era in which technology, industries, and nations have grown powerful enough to wage devastating wars—not just against each other, but against insects, on a scale that wasn't possible for humans of the past.[26]

In his essay "'Speaking of Annihilation': Mobilizing for War Against Human and Insect Enemies, 1914–1945," Edmund P. Russell III describes how war metaphors and technology were exchanged between the military and agricultural industries during the two world wars. The same language and metaphors used to describe the human enemy were also used for insects: they were to be *annihilated*. And agriculture was portrayed as a battlefield.

Once World War II ended, the technology of war was directed at insects too. Armed with new insecticides, which were often spinoff products of the poisonous gases developed to kill people, farmers and garden owners could wage their own war on insects. After thousands of years of insects being a nuisance which, to a degree, humans had to put up with, we could opt out of this coexistence.[27]

Are we about to win an irreversible victory in the war against our tormentors?

ANNE AND I were still standing by the rotten willow trunk. The two ladies and their poodles had walked on when we started talking about declining insect numbers.

The decaying wood and the wilderness around us in Svartdal was full of life. Contrary to what I expected, research shows that cities *can* be good habitats for many groups of insects and spiders.[28,29] And if we give them some basic consideration, many of them can thrive in our presence. They also thrive in parks and gardens left undisturbed by lawn mowers, and in areas where there are dead trees and shrubs. By contrast, the green areas around our cities are becoming more inhospitable. And the dividing and degradation of natural areas, the intensified use of agricultural land, and the use of increasingly destructive insecticides are some of the main reasons for it.[30]

What's good about insects is that they always show up. There are a lot of nature and wildlife experiences you can't get in the city. You'll never see a lynx, or beautiful rain forest lichen. But you'll definitely see insects. All they need is a neglected garden, an abandoned building site, or an overgrown embankment. On a spring day, if I crouch for a few minutes among the vegetables growing in the backyard, I'll find no end of ants, beetles, aphids, spiders, and wild bees.

WHEN I SAID GOODBYE to Anne at the edge of the wooded valley, I decided that I would put the spray bottle of poison in the back of the shed, unopened. I didn't want to participate in the war on insects. I could handle being reminded that the line between inside and outside was permeable. And I could handle being reminded that my backyard is also nature. A place where life forms much, much older than humans communicate and adapt.

When I got home that evening and sat, once again, at the kitchen table, I was struck by a sudden thought. The ants racing frenetically along the trail on my countertop were no less alive than me. Our human city, with its paved backyards and tiny gardens, was, from the ants' point of view, just another ecosystem they had conquered. To them it was perhaps *their* backyard. Did our presence there make more sense than theirs?

I still can't say that I like ants. They are too different from us, too hard to relate to. But I had gained some respect for these creatures: they are robust enough to conquer the world, and yet so finely tuned they can milk an aphid.

Instead of poison I put out some tiny bowls of sugar-free juice on the veranda, hoping the ants would take it back to their nest, and, believing it was sugar, feed it to the larvae and the queen. The colony would then starve to death, without any poison involved. It was a form of compromise between idealism and comfort.

After a week, the ants disappeared from the kitchen. This may have been due to the sugar-free juice. But it may have been because spring was fully underway and the ants had found other, more tempting things to eat. I had saved my relationship at least.

I can't say I miss seeing ants on the kitchen counter. But I kept a close eye on the ants and aphids in the backyard. And at some point in July, a new process began—the decaying process. One day the plant stems were covered in mummified aphid remains, empty shells with their butts aloft and their snouts embedded in the stem. A parasitic wasp, an insect related to the ant, had been at work. It had pierced the aphids' bodies with its long ovipositor and laid its eggs, which then hatched into hungry wasp larvae. The ants had been powerless to stop their lice being eaten alive, from within, by the parasitic wasp's offspring.

The Seagull Paradox

I F YOU TAKE THE FERRY southwest from Bodø, a Norwegian town just above the Arctic Circle, for a while you'll see nothing but gray water. Even a glorious early summer day will appear misty through the double-glazed windows. After forty-five minutes, flat strips of land will emerge on the horizon, which, as the boat approaches, develop a greenish hue. You might see gray-brown dots moving about on the boulders. But the trees, and the houses, which look like colorful building blocks scattered across the fields, remain invisible until the ferry lowers its speed and cruises into one of the straits between the islands. In these sheltered waters you might notice the sudden absence of swells. Eventually the boat heads toward a large red-painted dock, where the engine is taken out of gear and quickly put in reverse before the boat thuds against the tractor tires lining the pier. A large sign on the wall informs you that the dock serves as an offloading point for fishing boats and that you've arrived at the fishing islands of Fleinvær.

As I crossed the rickety bridge, which for a few minutes connects the boat with dry land, I noticed the gray-clad figure in a black hat waiting for me beside one of the piers. Magne Jensen is from Fleinvær and has lived on one of its islands, Langholmen, for all of his sixty-two years. He wears a ring in his ear, and listens to current affairs on the radio when out on his fishing boat. He knows the names of all of Fleinvær's 250 islands, and in winter he can navigate in the pitch dark between the numerous sandbanks without using a spotlight. He's the kind of guy who will put on an old Yes album at night, with his veranda door wide open, and let the music echo across the rocks.

Magne's little motorboat was tied up nearby. I climbed on board while trying to appear familiar with boats. He then started the engine, and I clung tightly to the gunwale. From the boat I could see the brown wrack line left by previous winter storms stretching far inland, slowly decomposing into soil. No matter where you are on the islands, the sea is never far away. As we sped out into the strait, I thought about how different life out there was. I'd never visited a place where the only means of escape is a ferry that comes just once every two days, and where no streetlamps disturb the starry sky. Where even the soil comes from the sea.

Although all the boats we passed were equipped with the latest technology, and I knew very well that Magne had a smartphone in his pocket, there was something timeless about the islands. As we passed boathouses and flat, mud-bottomed bays exposed by the low tide, Magne pointed out slight elevations, mounds of green grass that blended almost unnoticed into the landscape. "Burial mounds," he explained. These fishing islands have been inhabited since the early Iron Age and longer. Perhaps the bones of a local Viking chief rested under the green hills.

WE PULLED UP beside a huge boulder. The gray-brown dots from earlier stumbled toward us, grew legs, and turned out to be a flock of curious wild sheep.

I was there to survey the vegetation on one of the islands. It was the beginning of June and the orchids had just started to bloom. There was a lot to see, though vegetation isn't the point of this story. When we had finished crawling through the grass on the island, and the wild sheep had been given the stale bread from Magne's rucksack, we got back on board the little motorboat. I was to be given a tour of the other islands as we waited for the ferry that would take me back to the mainland.

A pair of terns hovered gracefully in mid-air above one of the white sandy beaches. Their screeching reminded me of something. Something that was missing. Magne slowed our speed and then stopped the engine. When the noise of the motor had stopped ringing in my ears, I noticed the silence. And once I was aware of it, it seemed a little strange—like someone had pressed "mute" on a movie I was watching.

Magne pointed at one of the islands. "The lesser black-backed gulls used to nest over there." There used to be thousands of birds. But over the last few years, the gull colonies had disappeared, one by one, he told me. Now just a few small ones are left. The brim of his black hat couldn't hide the sadness in his eyes. That year he hadn't seen a single lesser black-backed gull. "There are far fewer seagulls in general now. Herring gulls, lesser black-backed gulls, and common gulls," he said.

I was blown away. I knew that seabirds were struggling, but was it really that bad? Seagulls?

Magne started the motor again. As we headed back to the boathouses, he talked about the time when he was a boy, when the place had been teeming with people and seabirds. Seagulls had been as much a daily feature as the fishing boats and sheep. It was normal for seagull eggs to be on the menu. Now the island had few permanent residents or fishing boats out there, and Magne no longer foraged for seagull eggs.[1] He wanted the seagulls to come back.

I DIDN'T KNOW many people who wanted to have seagulls for neighbors. In fact, I knew a lot of people in the city who had grown to hate them. A friend working in nature management in one Norwegian town had told me that every spring and summer they are flooded with enquiries from people demanding that seagull nests be removed from their roofs. The newspapers were full of complaints about aggressive urban seagulls, and sometimes ran stories about people destroying the nests or killing seagull chicks. I know a lot of people who, in the heat of the moment, will declare how much they hate seagulls. But on the islands of Fleinvær, they were sorely missed.

I SAW THE SEAGULLS as soon as I walked through the glass doors of Oslo Central Station, on the very last leg of my journey home from northern Norway. They were guarding the kebab stand near the bus stop, dozens of them, and they were big. The sound of their screeching merged with the noise of trams, buses, and people. One of them jogged across the asphalt to become airborne, heavy as a fully loaded passenger plane. I moved out of its flight path. Once it was off the ground, I realized how big its wingspan was. If a seagull and I were to embrace, I'm not sure who would reach farther around the other's back.

A Dangerous Kindergarten

According to an expert on urban seagulls, Arild Breistøl, city-dwelling seagulls are a fairly new phenomenon. I gave Arild a call two days after returning from Fleinvær.

"The first herring gulls moved into the center of Bergen in the mid-nineties," Arild told me over the telephone, in a slightly faded southern accent. Bergen's harbor area, Bryggen, is known for its fish stalls and aquariums containing live fish. Fresh fish has been an attraction at this outdoor market for hundreds of years. Today the stalls are so overrun with seagulls it's hard to imagine it being any different. But it has been.

"I've seen photos of Bryggen taken in the 1930s. And there wasn't a single seagull!"

According to Arild, Norway's cities were more or less free of seagulls at the beginning of the 1900s. "Common gulls were the first to arrive, in the early 1990s," Arild said. Lesser black-backed gulls arrived soon after, and began nesting in the city during the late 1990s. At first, the numbers increased slowly, but they must have bred very successfully because in 2010 this increase really accelerated. The first herring gulls to nest in Bergen were registered in 2013. "If herring gulls follow the same pattern as the lesser black-backed gulls, we can expect their numbers to increase in the cities in the coming years."

The same pattern is seen in many European cities: in Bristol, in the United Kingdom, for instance, the first nesting seagulls were recorded in 1972, and in the last fifty years the population has grown to approximately 2,500 pairs. Cities are becoming increasingly important for gulls: in 2000, 15 percent of the herring gull nesting sites in the UK were in urban areas, and that number has probably increased since.[2] American gulls seem to be following suit: herring gulls, for instance, are nesting on buildings near the Great Lakes as well as in coastal cities like New York.

THE DECLINING POPULATIONS of gulls in the natural areas they came from is serious. Arild told me that the seagulls I had seen at Oslo Central Station could be seen as refugees. SEAPOP, a research program which has monitored Norwegian seabird populations for several decades, publishes reports that make gloomy reading for seabird lovers. Even common species like the herring gull are in steep decline.[3] In many places, the diminutive common gull is struggling even more.[4,5] Today, common gulls, which are the most numerous gulls in the cities, often end up being a nuisance for humans. But the high numbers in the city mask the fact that there are fewer of them where they come from. In just thirty years, half of the common gulls in southern and western parts of Norway have disappeared.[6] And since the situation looks grim in the rest of the country too, the common gull is on the national red list of endangered species. The days when common gulls could be found on most of Norway's islets and reefs are over.

"It's normal for seabird populations to fluctuate. But the prognosis looks bad for so many species, and has done so for such a long time, that it cannot be blamed on normal variations," Arild said. The problem is not limited to Norway. The herring gull is a species of conservation concern in the UK because its population has decreased by more than half in the last twenty-five years.[7]

WHAT HAPPENED? Well, a lot has changed along the coast over the last century, and it hasn't been good for seagulls. One thing, Arild explained, is that seagulls nest at the water's edge, on islets and reefs, where humans want to ride Jet Skis and build cabins. "Another important thing is the food," Arild explained to me over the phone. "Seagulls are bad divers. They live on what they find floating on the surface." I thought about this idea for a second. I had never seen a seagull underwater. They can't even stick their heads down and raise their tails like mallards do. They are too buoyant; they just bob about on the surface like they are made of paper.

Arild said that a few generations ago it would have been usual to see minke whales hunting for fish in the fjords of western Norway. Minke whales would chase shoals of fish up to the surface of the water from below. And in the chaotic aftermath of such an attack, seagulls could easily pluck the fish or what was left of them off the surface. On the islands of Fleinvær, Magne Jensen had talked about how puffins and razorbills would similarly gather up shoals of herring off the coast. These skilled swimmers can corral the fish into so-called herring balls and herd them up to the surface. On a calm day, Magne had told me, these herring balls seemed to be boiling with fish. There was so much herring in the sea that sometimes just a few hundred yards separated each herring ball. And because of the swarming gulls that gathered on the surface, these herring balls were often visible from far away.

According to Arild, fewer whales and diving seabirds are chasing food up to the surface these days. In the area south of Lofoten, the numbers of colorful puffins are rapidly declining. There are fewer razorbills, and minke whales are only occasional guests in the west Norwegian fjords. The fish remain in the depths, inaccessible to the gulls.

The small fishing boats that fished near land also provided seagulls with a lot of fishing waste, Arild explained. Most of today's fishing takes place far out at sea, on large self-contained trawlers. The restructuring of fisheries, and the generally far lower numbers of fish in the fjords and along the coast, means that fishing waste is no longer a food source for seagulls. The seagull's access to food on the coast is also very much affected by climate change. When the ocean's ecosystem is shifted, it creates winners and losers. The seagull, bound as it is to the ocean's surface, is one of the losers.

The Gaze of the Newcomers

The seagulls we saw in Vaterland, one of Oslo's most heavily built-up inner-city areas, didn't come across as losers. There were lots of them, and they were quarreling loudly about our bread crusts. The dirty-looking river flowed slowly under the bridge. I was meeting Bjørn Olav Tveit, a publisher and bird-watcher. When we spotted each other in the crowd, he gave the secret sign that all bird-watchers recognize—he revealed the binoculars he was carrying in his inside pocket. They were Swarovskis, the Mercedes of binoculars.

Bjørn Olav had jumped for joy when I asked if he wanted to join me on a seagull-spotting walk. We had hundreds of specimens of reviled urban nature before us. Bjørn Olav threw a piece of bread toward the river, hoping to attract the birds' attention. He wanted to demonstrate that seagulls are essentially cooperative animals.

"If you throw a slice of bread to a seagull, the first thing it will do is call to any seagulls that might be nearby: *Look, FOOD! Over here!*"

And sure enough, one of the seagulls squawked, and within a few seconds the air was full of wings and beaks. Classic seagull hysteria.

I'VE NEVER MADE much effort to learn how to tell seagulls apart. The ornithology books make it look easy: some gulls are dark, some have light feathers on their backs, some of them are big, others are small, and there's something about the colors on their beaks and legs. But in reality identifying gulls is far more difficult. Gauging their size from a distance is tricky. Their colors all blur into one. And, of course, young seagulls don't follow this pattern anyway; they are just speckled and rowdy whatever the species.

But Bjørn Olav was able to help me. He pointed left and right, explaining that the seven seagull species that nest in Norway can be roughly divided into two groups. Large gulls, which have powerful beaks and a line above the eye that gives them a slight frown. The herring gulls flying around us were typical large gulls, Bjørn Olav explained, with a beak powerful enough to crush a snail shell. I looked at the seagull he was pointing at, a large individual with stern eyebrows, and understood what he meant.

Then there were small gulls, and the bird about to get the piece of bread was one of them. It was clearly smaller than the herring gull. Its beak was too narrow to crush a snail shell but perhaps good for picking a snail *out* of its shell. It had rounder eyes that lacked the severe-looking eyebrows of the large gull, and so it had a more kind-natured demeanor. "A common gull," Bjørn Olav explained. The common gull sailed through the floundering mass of herring gulls and snapped up the piece of bread. The rest of the seagull mob launched a frenzied pursuit. So much for being cooperative. One of the pursuers swooped down to attack, and in the chaos, the herring gull dropped the bread into the river where it was snatched by a mallard. The seagulls sailed back to Earth dejectedly, and perched along the riverbank.

ONE REASON FOR seagulls moving into the city is associated with humankind's love of beautiful furs. In 1927, the American mink was introduced to the first Norwegian fur farm.

Since then, over the course of a century almost, numerous mink have escaped from these facilities. Today, there are wild mink populations in every region of the country.

Mink are mass murderers in fur, extremely efficient killing machines and egg eaters that will clean the table when they find a bird colony. Mink are also skilled swimmers and have no trouble reaching seabird nests on faraway islets and reefs. The birds have no way of defending themselves from their mink predators, which are, in an evolutionary sense, a completely new enemy. As a result, a nesting seagull colony raided by mink will often lose a whole year's production of eggs and chicks. It's not unusual for seagulls to abandon nesting areas that are vulnerable to mink.[8] Or as Arild put it, "Imagine everyone in your local kindergarten getting eaten, year after year. It wouldn't be strange if you wanted to look for a safer place to send your kids."

Flat roofs on tall buildings—preferably close to people—are mink proof. For example, according to Arild the roof of the Natural Science Building at the University of Bergen has become the municipality's largest bird cliff.[9]

I WONDERED WHY, if seagulls found life in the city so much easier, other endangered seabird species weren't moving to its flat roofs and the abundance of food? Terns, for example. Common tern populations are shrinking rapidly too. Their lives are similar to seagulls in many ways; they live on fish and food they find floating in the sea, and build their nests on islets and rocks along the coast. Tern eggs are also vulnerable to mink. Yet there are no terns living in Oslo. Why was it that only seagulls had taken over the cities? Arild had the answer: "Seagulls are just incredibly adaptable. While terns and many other sea-birds will only eat fish, seagulls can find food in the most unimaginable places."

When I think of seagull food, I think of garbage. And seagulls do a great job as garbage collectors. If there were no seagulls in the city, you'd get more rats, according to Arild. Take that, seagull haters! But garbage is often just a secondary source of nutrition for gulls. On the coast there are lugworms, which are evident from the coils of sand they create near the shore. Lugworms are an important food source for many gull species in coastal areas, said Arild. Even urban seagulls eat more worms than they do garbage. But what urban seagulls gorge themselves on are the worms living in our fertilized lawns. Roundabouts and traffic circles, grassy verges and borders, and other green areas can be full of earthworms, and these are the places seagulls will go looking for a bite to eat—almost as if they were plucking lugworms from between the piles of seaweed on a beach.

Humans have made it even easier for city gulls to get food. Thanks to urban street lighting gulls can pick at worms, and the leftover kebabs lying in the street or garbage cans day and night—making cities essentially like a twenty-four-hour buffet. Arild told me that he'd seen seagulls grazing in illuminated traffic roundabouts in the middle of the night.

He had also seen seagulls hunting larger prey. Arild told me about one of the lesser black-backed gulls he had tagged. "It always comes when someone feeds the birds. But it's not interested in the bread. It's after the sparrows."

Not only have I had to revise my opinion on crows, seagulls appear to be smarter than I'd originally thought as well. According to Arild, seagulls often flock to large open garbage dumps. Except on Sundays. "Somehow, they know the landfills are closed on Sundays, that there will be no more food, so there's no point in going there. Other seagulls will commute halfway across town to hang out at their favorite kebab shop, knowing what time the garbage will be emptied and that the owner won't bother chasing them away."

Because, like crows, seagulls have no problem telling humans apart. Many seagull lovers know a seagull that returns every year or that always begs for food outside the window. Arild told me about seagulls he had previously captured that consistently fled when he showed up. They must have recognized his face, just as crows associate certain experiences with humans. Friendships with seagulls can be lasting too. I have a friend who knows a herring gull, as in, he hand-feeds it every day. The bird calls out to him with a particular cry and immediately flies over when my friend opens the veranda door. The gull is not so enthusiastic if a different family member comes out.

According to Anouk Spelt, a behavioral ecologist at the University of Bristol, it's not just facial recognition. Seagulls also have a great sense of time: they are known to show up at specific schoolyards at lunch hour, just in time to grab the leftovers from kids' meals. Spelt says we shouldn't be surprised by the seagulls' awareness of time. She says that in natural areas, "Gulls know when to follow the tides in Bristol and now they are able to match the temporal patterns that we follow."[10]

I STARTED TRYING to look at the city around me through the eyes of a seagull. Here beside the river in central Oslo, where Bjørn Olav and I stood, the sun was streaming into tens of thousands of windows. There were no parks or gardens to be seen; the only visible green patches were grassy verges. Smooth facades grew skyward from the asphalt and carved up the surrounding airspace. I saw a landscape that was full of nesting sites, flat roofs high above the ground, inaccessible to two- and four-legged beasts without wings. Perhaps the discarded hot-dog buns lying in the street on a Sunday morning were the city's answer to the dead fish and marine creatures you might find on the beach after a storm?

Perhaps seagulls moving to the city is just logical, I thought. Perhaps the city gave them what they needed when their original habitat could no longer keep them alive. And this migration didn't just apply to the seagulls down by the river.

A SHORT DISTANCE from the bridge where we stood was a subway station. If the city was a body, perhaps the subway station wasn't so much its navel but a dirty and unmentionable place at its rear end, flushing out waves of busy-looking people. In the street outside, pigeons were busy peeling trampled matter from the asphalt in an effort to turn it into something edible.

The scenes playing out by the subway station could have been in almost any city in the world. The street pigeon, *Columba livia domestica*, has invaded the towns and cities of nearly every continent and is almost entirely dependent on humans. Fittingly, another globetrotter, the explorer Christopher Columbus, borrowed his name from the pigeon: the Latin word *columba* means "dove."[11]

Where do pigeons enter the picture? The terms "dove" and "pigeon" are used indiscriminately for smaller and larger members of the Columbidae bird family. But whereas the dove is a symbol of peace and beauty, the pigeon is better known as a pest in cities, even though the boundaries between doves and pigeons are blurry at best. The greatest difference seems to be how humans perceive them: if the birds are distant enough for us not to smell their shit, we call them doves. The lousy ones that eat our leftover buns, we call pigeons.

STREET PIGEONS ARE a variant of the species rock pigeon (or rock dove, if you wish) *Columba livia*. Before humans started building cities, rock pigeons lived on seeds and nuts in open, arid parts of Europe, North Africa, and East Asia. The rock pigeon's descendants spread with humans to cities all around the world where they were used as carrier pigeons, among other things. But these pigeons escaped, or were released. And they were very well suited to living freely in the cities humans built, for at least two reasons.

First, city pigeons often nest in tall, vertical structures—on ledges and bridge spans, for example—which offer the same level of protection from predators as the cliff faces they came from. For pigeons, cities perhaps represent some kind of strange canyon landscape.

Second, pigeons also have a kind of super-skill: they can live on an all-plant diet, even in their youth. Very few bird species have young that are vegetarians. To grow, most chicks need meat or insects, a throwback to the carnivorous dinosaurs they are descended from.[12] But pigeons do not need meat. Pigeon parents use the same strategy that mammals use to give their young a fat- and protein-rich diet: they produce milk. Adult pigeons, both males and females, produce a milky fluid in their crop, an enlarged part of the esophagus that stores and softens food that they regurgitate for their young.[13] This self-reliance allows pigeons to raise young in even the most insect-free urban canyon landscapes.

Pigeons essentially behave a bit like human city-dwellers, I thought, as Bjørn Olav and I stood by the subway station watching them pecking at a bag of donuts from a nearby kiosk. In other words, they live very close to each other in high-rise locations, prefer to behave as a group, and eat more or less the same things humans do—a lot of junk food and fast carbs.

ON THIS JUNE DAY, dark eyes could have been watching us from the tallest of the high-rise buildings that loom over the area. That's because an old enemy of the pigeon has moved from the rocky wilderness to the city as well. Oslo Plaza is currently Norway's second-tallest building, and it's the place in Oslo where you are most likely to see peregrine falcons. These falcons like to watch for prey from on high. And from the top of Oslo Plaza, falcons can reach their maximum speed, 240 miles per hour, on their way down to scoop up a street pigeon.

Peregrine falcons aren't the street pigeon's only problem. The beams supporting the bridge where we stood were covered in bird spikes: steel pins that look like supersized barbed wire and make it impossible for pigeons to land. Pigeons are, at the very least, just as hated as seagulls. We have built pigeons the perfect habitat, and even helped them to get there. But we have given them an ambivalent reception.

URBANIZATION OFTEN MEANS CONFORMITY. Just as you can find the same green Granny Smith apples in the fruit section of supermarkets around the world, or the same chain restaurants in the streets of every major city, you can share a café table with a street pigeon and a sparrow, whether you are in Moscow or Odense, Delhi or New York. But there are exceptions: the artificial rock faces of the city are used by other, more infrequent guests, and Bjørn Olav and I found one of them in the middle of Oslo's former banking district.

The tram rumbled up the street as we searched for the right entrance. For some time, the area had been notorious for drug dealing. The drug users had been forced elsewhere, and the building before us now contained a fancy news agency. But we weren't looking for news or drugs. We were on the trail of one of Norway's rarest birds.

We pressed the button on the intercom and were then buzzed into the backyard by Marte from the news agency. The yard was a rectangular desert of stone with weeds poking up through the crumbling asphalt, surrounded on all sides by a five-story apartment building. It looked like the exercise yard of a maximum-security prison.

Had Marte not pointed them out, we would never have noticed the three-foot-long slits in the masonry. They were about four inches wide and twenty inches deep, and they stretched all the way up the building. These slits, which might once have seemed like a smart idea by the architect, had become the site of an ornithological sensation: every year, two or three black redstarts come here and nest. A little bit of Google research told me that the black redstart is a pretty little member of the thrush family with a black breast and a rust-red tail that flicks when it walks. And the only place in Norway where this bird nests regularly is in Oslo's most congested inner-city area.

What is this rare bird doing in the middle of one of Oslo's least green streets? In more southerly parts of Europe, the black redstart is a mountain bird, an expert at living in remote, sparsely vegetated areas where it nests in rocky crevices. There it has adapted to surviving on the few insects that fly around in the mountains. In the city, therefore, it can survive on the few insects that live on the yarrow growing between the tram tracks and the municipal flower boxes filled with imported violets. Perhaps it sees the cracks in brick buildings as cracks in the mountainside and the built-up streets as similar to the ravines in the rocky landscape it comes from.

Along with the rock pigeon and the peregrine falcon, the black redstart is part of a wild ecosystem that has moved into the inhospitable heart of the city.

A BIRD-WATCHER CANNOT have slow reflexes, nor can they be offended when being interrupted in mid-sentence. I had to work a bit on the latter because in the middle of a conversation Bjørn Olav would suddenly whip out the binoculars, raise them to his eyes, and call out the name of some species or other.

After leaving the cliffs of the city center, we passed through the large wrought-iron gates of the botanical garden a little farther east. We followed the path beneath pale-green willow trees and walked beside gardens of low-growing alpine flowers. A constant hum in the tall grass revealed the presence of insect traffic.

Soon, a noise that sounded like a fingernail scraping the edge of a comb caught Bjørn Olav's attention. A split second later, his binoculars were at his eyes. "Greenfinch!" he called out. From the sound I had actually guessed that it was a greenfinch, and felt very pleased with myself. A flock of greenish birds flew from one tree to another.

The contrast to the hard streets of the city center was enormous. There wasn't a pigeon or seagull in sight. I stood there absorbing the sounds and smells of the garden. It was a nice place to be; it somehow felt more real than the urban landscape we had just left. It was strange to think that this landscape has to be constantly maintained by weeding, planting, and tending. The park, and its birds, is in fact no more natural than the seagulls, pigeons, and peregrine falcons in the city center.

About a hundred bird species have been registered in the botanical garden, an almost unbelievable number for a little green patch in the middle of a city.[14] Contrary to what I'd expected when I began researching life in cities, both the variety and density of birds can be *higher* in cities than in many other habitats. Some bird species can be abundant in the city, even when they are otherwise in decline.[15] One such bird is the starling.

Bjørn Olav pointed at a dark-brown bird walking confidently on the lawn beside us. The starling, a medium-sized bird with dark feathers and a yellow, upward-pointing beak, can look quite similar to a blackbird. But the starling's summer plumage is a shimmery blue-green flecked with white like a bejeweled ball gown. It is like a blackbird dressed for the opera. I also noticed how differently the two birds move. The starling walks or strides like a human, unlike the blackbird which hops on two legs.

Starlings were previously found all over the rural landscape in its home range in temperate Europe, but they are now on both the Norwegian and the English red list of endangered species. In the UK almost 70 percent of the rural starling population, and 90 percent of those that lived in the forest, have disappeared over the past fifty-five years. Like seagulls, starlings have found conditions are often better for them in the city. A study conducted in Sheffield, England, for example, shows that starlings are far more common in urban areas than in rural ones: urban areas cover just 7 percent of the country but are home to 40 percent of the country's total starling population. The authors of the study write that urban areas can serve as refuges for many bird species that can no longer live in intensively farmed rural areas.[16]

In North America, starlings also flock to cities. In his brilliant book *Darwin Comes to Town: How the Urban Jungle Drives Evolution*,[17] Menno Schilthuizen tells the story of how starlings first colonized cities in the New World when an amateur ornithologist and (more importantly) Shakespeare lover released 100 birds in New York City's Central Park. Though the details have been embellished somewhat over the years, it's still believed that all of North America's approximately 200 million starlings are descended from that first flock.[18]

BJØRN OLAV EXPLAINED that the urban migration of some bird species has a lot to do with insects, which birds need to raise their young.

"Insects are the key!" said Bjørn Olav. "If a place attracts insects, the birds will come. In Norway, pesticides are less used in cities compared with the surrounding agricultural areas, so cities often provide better conditions for insects."

As we walked through the park and peered into the trees, I thought about why I had suddenly become so hung up on city birds. On the bookshelf at the house where I grew up was a three-volume set of collected folktales that was inherited from some distant relative. It was a sumptuous edition, bound in red leather with gold edging and gold lettering on the cover. The pages were delicate, thin as tissue paper and with a sweet smell. These folktales depicted a time when people were used to dealing with wolves, bears, and lynx (in addition to trolls and smart-mouthed princesses), and the people in those tales treated these big, wild animals like old acquaintances, giving them nicknames and personalities.

This intimacy with wild animals was an alien concept to me because the animals in these books are no longer a part of everyday life. They lost the battle over their habitat and, for most of us, now belong to a fairytale-like past.

Today, other more familiar organisms are disappearing. Between 1980 and 2014, Europe's bird populations declined by 20 percent.[19] The situation is especially bad for ground-nesting birds, with over 70 percent of European species in persistent decline.[20] The situation is not unique to Europe: a study from 2019 showed that about 3 billion breeding birds, or 30 percent, have been lost from

North America in fifty years.[21] In Norway, the numbers of many common birds—such as great tits, sparrows, and greenfinches—are also going down.[22] Will seagulls and crows soon be the only birds making noise in our cities? Will our coastlines, mountains, and cultural landscape become even quieter?

Perhaps these thoughts had led me to become so absorbed by city birds. Possibly because that's how the human brain—at least the one I have access to, namely my own—works. The thing we are in danger of losing suddenly becomes interesting.

BJØRN OLAV AND I left the birdsong in the botanical garden behind and followed the streets west until we reached Oslo's city hall. The city hall is located right by the seafront and has two tall towers on either side. From a distance the building looks like a huge musical instrument designed to be played by the winds from the north or the sea breezes from the south.

I could appreciate how, to a bird, the tall buildings might resemble cliff faces while the flat roofs might look like islets and reefs in an ocean of asphalt.

At the top of a wide staircase leading to the city hall's entrance, a fountain sprayed water into the air. Bjørn Olav and I stood looking at a street pigeon as it drank from a nearby puddle. The pigeon raised its head and let the water run into its throat as its metallic pink neck sparkled in the sun. Seagulls walked around the pigeon flock.

The seagulls and pigeons around us were as much a part of the city as the concrete and the traffic. Once exiles from completely different places, they now gather in landscapes that have never existed on Earth before. I practiced sorting the gulls into some kind of order, dividing them into big gulls and little gulls. It felt good to get an overview, to give them names and order them by species. They went from being a flailing gray-and-white mass to individual species that I could know something about. "Hello, little common gull! Hello, angry herring gull!" Perhaps they would eventually get new species names too, names that didn't refer to their past. Street gull? Burger gull?

WHILE KEEPING AN EYE on a lesser black-backed gull (if it was going to attack a pigeon, I didn't want to miss seeing it), I thought about why gulls and pigeons are despised so much. Perhaps their great sin is that they didn't remain in the world of "wild" nature. Instead of gradually dying out as the pristine nature around them deteriorates, they have adapted to humans.

In her book *Crow Planet*, Lyanda Lynn Haupt writes that crows are the birds that we deserve. Ungainly and black, a reminder of our ecological mistakes.[23] I would say the same is true, perhaps to an even greater extent, of seagulls. Seagulls have fled the coasts for the city because of humans and our actions. At some level they annoy us more because they force us to confront our actions and behaviors when we witness the spectacle of seagulls engaging in dive-bombing raids and plucking food from garbage bins and takeout boxes.

The Ghosts
of the City

A teardrop-shaped seed, attached by a stem to a narrow, curved leaf. An exclamation mark under a light green comma. The wind takes hold, tears it loose from the tree it is hanging from, and lifts it high into the air. As it descends, the comma spins around, lopsided as it is. The seed floats several hundred yards from the mother tree. It lands in an open forest, in the shade of tall grass.

A CRASH WAS INEVITABLE. Until this point, exploring urban nature had been too much fun. My obsession with urban crows and ants had been far more powerful than my longing for the countryside. I'd been arriving late for appointments after being distracted by seagulls pulling up worms, and getting up in the middle of the night to listen to blackbirds sing. Perhaps this intense fascination was too good to last. But I hadn't expected this disillusionment to come from an encounter with my old friends, the trees.

It was mid-July, and nothing about central Oslo was wild. My thighs stuck to the bench I was sitting on, and the hedge surrounding the bench was dry and withered. It was one of those days when you might see heat haze on the road, when the city no longer smells like itself but like dust and rotting food waste. Other than the crows, which sat gasping in the shade, and the occasional pedestrian, there was very little life to be seen.

The hedge around the small square looked as woebegone as the crows. It was less than three feet high and planted in a horseshoe shape around two benches on a tiny patch of grass. Large dark-green leaves hung down toward the asphalt like ragged handkerchiefs.

I hadn't planned on leaving the city that summer. But right then I was ready to go anywhere, preferably to the forest for its shade and humid air. Even the hedge seemed lifeless, I thought, before a bumblebee buzzed past and into the foliage. I watched as it zigzagged in and out among the leaves, landed heavily in a cluster of yellow-green flowers, and then rummaged about upside down for a while. It was long enough for me to study the flowers. They were hanging below a curved, narrow leaf that resembled a light green comma.

Only then did I notice the smell: an intense fragrance with hints of orange. It reminded me of something—a July day a couple of years earlier.

That day we had been searching for really old linden forest. The "we" in this context was me and Per Marstad, author of several books about fungi and a legend in Norway's mushrooming community. Per is pushing eighty. He is a former senior men's weightlifting world champion and moves through the woods like an elephant (it takes a little while, and makes a lot of noise, but he is unstoppable). He has steel-gray and perpetually well-groomed hair. And he will share stories about his seafaring life if you ask him nicely; I had brought some buns with me for lunch and hoped they would do the trick. We were in an old south-facing broadleaf forest a few miles west of Oslo searching for rare mushrooms. When August approaches, mushroom lovers are longing for the forest.

As we walked through the mixed woodland, I noticed that it was getting darker. We passed hunched spruces with branches that drooped like heavy overcoats toward the ground. The terrain got steeper, and moving between the rocks became difficult. The lindens on this escarpment looked old. Huge root systems spread all over the rocks like a mass of tangled limbs.[1] And twisting up from these roots were great trunks full of knots and burls. The bodies of these ancient linden trees looked like something soft and alive moving in extreme slow motion.

The air in the forest was cool and humid. The dense canopy blocked out the sun and kept the moisture in. Trees had fallen and brought down others that had got in the way. And where these old trees once stood, light shone through gaps in the foliage onto carpets of dark-green liverwort leaves, reminding me that outside the forest it was a bright sunny day.

These fallen giants created new and tangled topographies. The trunks, which lay strewn all over the place, had been taken over by other forms of life. The decaying wood was riddled with holes and tunnels. The fleshy conks of bracket fungi stuck out from beneath the trunks, like mini flying saucers that had crashed into the bark. As I walked around, my shoe sank into the ground with a faint squeak. I had stepped on a piece of crumbling wood that was crisscrossed with white threads like a piece of avant-garde lacework. Tree death is no orderly process.

The trees creaked in the wind, while a woodpecker drummed nearby, its persistent knocking seemingly coming from all directions. Insects buzzed above our heads. The hum of individuals purposefully going about their business made the linden forest feel as busy as a major city.

I grabbed myself a leaf from one of the fallen trees. It was large, dark green, and heart-shaped, with rust-brown hairs on the back. The clusters of linden flowers had a faint citrusy smell.

I PICKED A LEAF from the hedge by the bench. This soft leaf was also heart-shaped, and its underside was covered with rust-brown hairs. My nose had not been mistaken. The little hedge *was* a small-leaved linden, *Tilia cordata*, trimmed and transformed into something unrecognizable.

Although I hadn't recognized the hedge as a linden, I happened to know a bit about these trees. The thing with trees in the *Tilia* genus are that they come close to immortal.[2] When a trunk keels over, a new trunk can emerge from the roots. Not only that—new roots, and new trunks, can grow from other parts of the fallen trunk.

And so, some of the linden trees in today's European forests can be several thousand years old, perhaps as old as the linden forest itself.[3]

These old forests contain an abundance of life. Below ground are rare fungi that love the stability and continuity of these giant lindens. And every now and then they break the surface, as many mushrooms do. It was these mushrooms that Per and I had been looking for that summer day in the forest.

THE LINDEN HEDGE wasn't the first piece of urban greenery I'd failed to recognize. One day in early spring I had gone for a walk in the neighborhood with a tree guidebook in hand. And that ramble turned out to be a disaster for my self-esteem as a botanist. I had started along the river, where huge old willow trees lean out over the water. I knew there were several willow species in Norway, and I'd had some training outside the city in how to tell them apart. But many of the willow trees in town were not in the book, perhaps because they were species imported from elsewhere. And many of the other trees were still leafless. So even armed with a guidebook, I was unable to determine the species of half the trees I found.

Perhaps it was a sense that the greenery of the city is an inferior version of the forest that had made me overlook the urban trees. But whatever these urban willows were, I was going to tackle their identification.

I got hold of a guidebook that also covered imported trees and started to take it everywhere. Whenever I passed a tree I didn't recognize, I would stop and check what the young buds looked like and how the branches grew. And after months of going round with a loupe and a guide-book constantly in my bag, that looking started to pay off. The neighborhood trees went from being an anonymous green mass to having individual names. Mountain ash and wild cherry. Daphne willow and gray willow. The beautiful tree with thin pale leaves hanging over the river turned out to be a white willow.

I REALIZED THAT learning the name of a tree was like viewing it through a lens. It defined, reinforced, and made the details sharper. Naming the trees helped me to *see* them. I noticed how the smooth bark of a young aspen—gray with a hint of green—was a completely different shade of gray from the gray-brown of a gnarled oak. I learned to recognize the narrow trunk of a young mountain ash, and the coarse bark of a more mature willow. Learning a few different tree species helped me to see the trees in a very concrete way: in the spring I discovered lots of trees in the area around my block that I'd never noticed before. They were bedraggled city trees confined to their own squares of soil. Each one was a living being with its own unique bark structure, smell, and flower, and most originated from other parts of the world: honey locust, sycamore maple, Japanese cherry.

My linden hedge seemed different once it had a name. Trees are given species names by humans, and these names often reflect the human world. For example, humans have used the flowers, leaves, and bark of the linden tree in various forms as a traditional remedy or pain reliever, and the name linden is related to the Norwegian expression *å lindre*, which means to heal.[4]

The linden tree has also been prized in many parts of the world for its properties. Its white inner bark, called bast fibers, consists of long cells with thickened walls. This small morphological detail makes the fibers strong lengthwise, much like cotton fibers to which they are closely related. And this detail has proved to be extremely useful. In North America, the bast was used to make cords and ropes, leading to the name basswood. In Norway, linden bast is used to make fishing nets and nautical ropes.[5]

As I sat on the bench beside the hedge, I attempted to break off a twig. But it was stubborn and stringy and wouldn't snap. I tried twisting it around a few times, without success, and eventually gave up and let it hang there.

The Ghosts of the City

In the city, the heat was as intense as ever. I lay down on the small patch of grass that the linden hedge kept hidden from passersby, and I pondered.

I knew that more than a quarter of all the trees in Oslo's inner city are lindens.[6] But what was it about the linden tree that made it such a favorite of urban landscape architects?

One reason is that unlike most other trees, the linden (like most of the other species in the *Tilia* genus) can be repeatedly cut, shaped, and pruned until it has the shape of a four-foot-tall square hedge, for example.

As I lay on the grass, I remembered how entomologist Anne Sverdrup-Thygeson had been merciless about Oslo's lawns on our insect-spotting trip earlier that spring.

A lawn, according to Anne, is a system that is kept at an artificial standstill. Maintaining this standstill requires constant effort, weeding, and/or pesticides to kill anything that might compete with the lawn grass. Lawn-grass flowers are wind-pollinated and are unable to offer insects pollen or nectar.[7]

But lawns are nice to lie on. While I lay there on the recently mowed grass, I thought about how grass manages to grow, over and over again, even when it is constantly being cut.

When grasses were evolving into what they are today, there were no lawn mowers. There were grazing animals. While studying biology I learned that plants have (at least) two main strategies for surviving in areas populated by hungry animals that want to eat them. One is to avoid being eaten, by developing thorns or toxins to make themselves inedible, for example. The other is to teach themselves how to live with being eaten, or even benefit from it.[8]

Plants that adapt to being grazed on often have several growing points, places where they can start growing again if their tops get chewed off. These growing points are typically found low down on the plant, so that it can keep growing even if a hungry animal gnaws it down almost completely.

In contrast to most other plants, for instance tulips, grasses are typical grazing plants. They can be gnawed right down to the root and then grow back with a new stem and flower. If a hungry deer happens to gnaw on a tulip one fine spring day, the stem will never produce a new flower. This ability to regenerate enables grasses to survive as their competitors struggle, and grazing animals ultimately help rid grasses of any troublesome competition for space and light.

As I lay in the short cut grass thinking about this idea, I was struck by the following question: Did the linden tree also adapt to being gnawed down and growing back? And if so, who or what gnaws on enormous linden trees? It had to be something very big.

I GOT UP from the lawn and went home. On my bookshelf I found *Arts of Living on a Damaged Planet: Ghosts and Monsters of the Anthropocene*, an anthology I had read in a totally different context a few months earlier. Its contributors are humanists and scientists who write about, among other things, how landscapes can be haunted by previous forms of life. One sentence had particularly caught my attention: "Northern trees that grow back when cut down . . . may have evolved that ability in times when elephants trampled them."[9]

That one sentence stood on its own, without further references, but it was enough to fire up my imagination. Because linden trees *have* lived in close proximity to elephants and other beasts.

The vast forests that covered the European mainland during the warm interglacial periods were once roamed by herds of a now-extinct elephant species. The straight-tusked elephant, *Elephas antiquus*, which weighed somewhere around fourteen short tons and had a shoulder height of thirteen feet, was one of the largest elephants that ever lived. It was twice the size of today's African elephants. With its flexible trunk it could reach leaves growing twenty-six feet from the ground.[10] The straight-tusked elephant was one of several huge, now-extinct mammals that grazed in Europe's forests.[11]

So, the linden family existed in Europe for long periods at the same time that straight-tusked elephants and other large mammals were roaming around.[12] Trees that willingly sprout again from their roots after being felled tolerate being grazed on better than species that are unable to grow back.[13] Could the linden tree's ability to grow from almost any point be an adaptation it made due to being grazed on by these huge animals? Could lindens, like grass, have adapted to being eaten and trampled?

I had talked to several knowledgeable biologists in this field, including Jens-Christian Svenning, who has carried out a number of studies on European megafauna; Charlotte Sletten Bjorå, associate professor of botany at the Botanical Museum at the University of Oslo; and my former professor of biology, the all-knowing Klaus Høiland. None of them could *confirm* that a linden's ability to withstand constant pruning is an evolutionary nod from the elephant. But they all thought it was a credible hypothesis and had no objections to it.

A FEW WEEKS LATER I walked past the bench next to what was by then an almost-flowerless linden hedge, and I sat down. It was an evening in early August and a cool calm had settled over the streets. I reached into the hedge and plucked what had once been a cluster of flowers. The thing dangling beneath the pale green stipule was about to turn into little green nuts, and its slanted, heart-shaped leaves were still soft. Hidden from view as it was by the low hedge, the bench was actually a good place to sit.

I closed my eyes and imagined what would happen if the hedge were just left to grow. Perhaps a shoot would rocket upward, grow taller, grow thicker, and eventually sprout branches. And the sky above this little square would soon be obscured by the tree's foliage. Perhaps the branches would smash a few windows in the brick buildings as they grew. Their roots would destroy the asphalt and, in search of water, penetrate the thick water pipes buried underneath. Then centuries, and perhaps millennia, might pass. Did the landscape architects who specifically chose linden for this spot consider the tree's long-term potential?

Today, Europe's megafauna is extinct. Giant straight-tusked elephants no longer graze on massive linden trees. Humans were too good at hunting them, and the climate became too hostile for them. Now humans and our hedge trimmers are the grazing animals. But the ghosts of these elephants are perhaps still here, in the form of a neatly trimmed linden hedge.

Maybe I've gone wildly off track here. But I like the hypothesis.

AS I SAT BESIDE the hedge I was still yearning for the forest. It would be lying to say otherwise. The city trees and the linden hedges, they're not *forests*. Forests don't happen in a day, a week, or a year. They exist on a different time-scale to humans. The linden forest is one point in a long story. But even a ragged little city hedge has its history.

Stories From the Underground

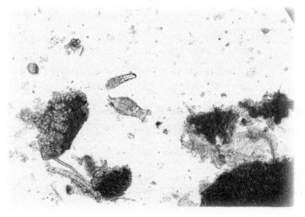

A rotifer maneuvers in a storm of humus.

You are the size of a pinhead and stumble into a tiny wormhole. As you fall through the layers of earth, the colors around you change from dark brown to light gray. The tunnel then curves and levels off, and you land with a thud on the moist, sticky worm poop covering the tunnel's interior. The air is humid and cool. You stand there with water up to your knees and notice something tickling your legs. The bristly fan-like mouth of a rotifer the size of a mouse is currently examining whether your trouser legs are edible.

Dark bodies of water, held in place by surface tension, bulge from the holes in the narrow side tunnels like gigantic black pearls. In the darkness you catch a glimpse of an earthworm the size of a blue whale blocking the passage with its distended body. A moss mite the size of a mammoth peeps out from a side tunnel. With its eight legs and claw-like mouthparts, it attacks one of the fungal hyphae stretching between the walls like thick drainage pipes.

In one of the convex bodies of water you see a transparent roundworm the size of a boa constrictor. Its twitching snout, shaped like a wreath of ostrich feathers, searches around for bacteria. The articulated body of a springtail, the size of a Labrador, collides with a wall and showers the tunnel with bacteria the size of marbles.

STOOD WITH HALF MY BODY in a hedge while struggling to dig a hole in the ground below. The hole was the size of half a basketball, and I had spent longer than expected hacking away at it. Just a short distance away was a footpath, where people were heading toward the subway, and I was waiting for someone to come over and ask what I was doing. The hedge stood on a tiny rectangular patch of earth, surrounded on all sides by asphalt. A bus thundered past on the nearby ring road and its sound echoed off the tall brick buildings that surrounded the open square I was in. Some wilted dandelion leaves dangled in the shade under the hedge. And two brand-new pairs of white underpants lay on the asphalt, the third was already in the hole. I have planted a lot of strange things in my life, but this one really took the cake.

It was a cool August morning and the city was shrouded in mist. I'd actually been out for some time already; before coming to this inner-city hedge, I'd gone and dug matching holes in an area of deciduous forest just east of the city. With a shovel in hand, I had taken the subway to a stop out in the suburbs, then marched against the flow of commuters, through the congested subway station, into an area of woodland full of birch and speckled alder. Between the pale birch-tree trunks I dug three holes and placed a pair of underpants in each one. The earth in the forest had been dark and tacky; small clods of it had clung to the tangled layer of grass roots. The earth I was toiling with now was hard, and crumbled into sand between my fingers.

The earth in the forest had had a couple of worms in each shovelful, and it smelled fresh. But there were no worms or any other living things visible in this urban soil. Life begins in the soil, they say. But the soil piled up beside my hole didn't seem capable of giving birth to anything living.

I looked down at the hole. Then I started digging another hole. All three pairs of underpants were going down into the earth. Although, as I continued to struggle, I thought about how the sandy mass I was hacking away at could hardly be called earth.

"Earth" is such a big word. It can be anything from the handful of dirt you sprinkle over your recently sown seeds to the entire planet we are living on. Earth was something I purchased in big bags and shoveled into pots for my indoor plants or into the raised flower beds in the backyard. It was an airy, black mass that gave off very little odor. But now and then it would go off; it would attract flies, or a white deposit would appear in the pot. I would then throw all that earth in the trash and buy some more. I had no idea how to make the sick earth healthy again.

I knew that earth is related to mold. The word "mold," however, has something musty about it and makes me think of black-clad gravediggers. Although it was also possible that mold has something to do with humus, but what exactly is humus?

The sandy matter crunching against my spade certainly wasn't mold. It was totally different from what you get in the sacks from the garden center. Something told me that it wasn't suitable for potting, that tomatoes would not grow well in it. But if it wasn't earth, what was it? Sand? Dirt? My vocabulary betrayed me; I had no suitable words.

People carrying yoga mats and laptop bags gave me puzzled looks as they strode by on the sidewalk. Had they never seen an ordinary citizen digging a hole before? Perhaps not. I had never seen anyone who was interested in city soil either. Except for children. Kids are always curious about what's under the ground. I'd seen children lifting up stones with their friends and squealing in both horror and delight at what was going on down there. Kids also bury things: dead birds and pets, laid to rest with that undefinable sense of ceremony that children have.

I'd been interested in plants for many years, but only the part above ground. While leaning on my shovel, I thought about how the city's soil is hidden beneath the asphalt and paving stones, and below the dense and heavily trimmed grass in parks. I hadn't buried anything for a long time, or lifted stones to see what was under them. I realized that I might have known more about soil—about its living creatures, taste, and smell—when I was three years old than I do today.

I put one of the underpants into the hole, about five inches deep, and shoveled the earth onto the white cotton. They were just a cheap pair bought half-price at H&M; nevertheless, there was something ritualistic about watching the white material vanish into the darkness.

The Sound of Compost

I had, however, one other entry point to the mysteries of urban soil: a three-foot-high green barrel, hidden round a corner in my backyard, partly obscured by a hedge (and out of sight of my neighbors). The barrel is wider in the middle and has a lid. This old hot-composting bin was something I'd inherited a few months earlier, and I found myself suddenly forced to take an interest in the journey that food waste makes—from the bucket under the kitchen sink into fertile soil.

But this journey from food to soil was a lot more difficult than I'd expected it to be. When the compost process is at its most effective, the pile of decaying organic matter is a steaming mass that will swallow bucketloads of potato peels and stale bread without gaining an inch. Every time I open the compost bin—after emptying a bucket of food into it the day before—I'm equally surprised. Hundreds of pounds of food waste can, in a few months and in favorable conditions, be transformed into a small barrow-load of compost soil.[1] This soil is black, like the soil available at garden centers. But whereas the soil from the garden center is almost odor-free and homogeneous, compost soil has a fresh smell and lumps that feel greasy between your fingers. If I shovel it onto the vegetables in the backyard, the cabbages will literally sprout after a few days.

But should some change in the balance between moist and dry, carbon and nitrogen, alter the equilibrium between the microorganisms, in just a few days the mass can go from an efficient composting machine to a stinking swamp that's crawling with maggots. You may have been caring for it as diligently as possible, adding wood shavings for texture and feeding it regularly. Nevertheless, it will begin to smell like vomit or manure (which is why I hide it from the neighbors). It takes a lot of effort to get this rancid soup in better shape. Which is why I love my compost when it is perfectly balanced.

If the compost is steaming happily when I open the lid, and yesterday's onion skins are already unrecognizable, I feel good. I get a sense of things moving forward, that someone is working on my team. In contrast, when the compost is out of whack I feel deeply uneasy. It's frustrating in a way that's similar to a mild version of being stuck in a traffic jam.

I have learned that too much salt in the compost will ruin the decomposition process. And that banana peels containing pesticides should not go into the barrel. Washing the compost bin with chlorine is also a bad idea. But other than that, whatever was happening in the compost bin remained a mystery. What is it that breaks down eggshells and bread crusts and turns them into this rich black mass? And when the compost is shoveled onto the garden beds, what happens then?

The more I thought about composting, the less I understood.

I STARTED TO LOOK for exposed urban soil on my way to the bus and the subway. Patches of it stuck out here and there, back from its exile beneath the asphalt and concrete, on roundabouts and grassy verges. The trees along the street were parked in their individual soil squares. And these individual patches all had their own character. Some were soft and black; others were sandy and gray. Some had plants growing in them; others were barren. And some were just full of cigarette butts, broken glass, and other foreign objects. I began digging up the soil and looking at it under my loupe. But even at twenty times magnification, all I could see were brown and white grains of sand and perhaps a small amount of black mass. Was this soil alive? Could plants grow in it? And where had it come from?

I HAD LIVED all my life in the city without ever thinking about the soil I was walking on. That's the reason I was there, on that August morning, patting down the soil over the last pair of cotton underpants and marking where I'd buried them with a popsicle stick.

I KNEW, IN THEORY, a little bit about soil. I knew that well-functioning soil is fundamental. It is so essential to human life and well-being that a number of cases of societal collapse—the Viking colony in Greenland, for example—have been linked to soil degradation.[2,3] Soil captures rainwater and lessens the risk of catastrophic flooding. It absorbs rain like a sponge and stores it to prevent plants from drying out. And it decomposes anything that needs to be broken down. People who work in cemeteries where there is limited potential for expansion, in a city for example, are very interested in the soil's ability to turn organic matter into earth as quickly as possible. It makes their jobs far more pleasant.

But as I walked home, with the shovel slung over my shoulder, I realized I had only vague ideas about how this transformation actually happened. Or if it even happens *at all* in the patches of earth squeezed in between the asphalt and the paving stones.

And it's not just me who considers the life of soil to be unexplored territory. If you're sad that there are no longer any uncharted spots on the map still to be explored, just look down. Soil life has been described as "the third biotic frontier," second only to the unexplored life in the deepest parts of the ocean and the treetops of the equatorial rain forests. There are more microbes in a teaspoon of healthy soil than there are humans on the planet. Only a small number of the living things found in soil have been identified. But that small number makes up a quarter of all known life. In other words, if I wanted to reach unexplored urban wilderness, all I had to do was dig.

But these kinds of figures didn't help me. My brain perceives big numbers as simply numbers, not as images that I can connect to further my understanding of something. In addition, little research has been done on urban soil despite the fact that cities occupy increasingly large parts of Earth's surface. The first scientific articles on urban soil didn't start to appear until the 1990s.[4]

"THE BAD NEWS is that it's almost impossible to find a complete soil profile in cities," said Linda Jolly on the phone from NMBU, the Norwegian University of Life Sciences, where she works as a lecturer. "The original soil is long gone. It's a terrible place to establish a garden."

I'd called Linda, who is an authority on school gardens, on the day of my momentous underpants burial. Linda is a biologist and educator, and has been involved with countless school gardens since starting her first one in 1979. Since then, she has worked continually with soil, often within the city limits. However, urban soil often has to be removed, she said, because it's not fertile enough to support plant roots. She was explaining the connection between soil and compost. "When a school garden is being planted in a city, and we excavate the soil, it will often smell sour.

"If we're going to grow something, we nearly always have to bring in soil from somewhere else. It's an admission of failure. Soil should smell fresh, like soil in a forest. Not sour!"

The sour smell occurs when a lack of air entering the soil prevents decomposition from occurring normally. When that happens, the soil gets taken over by anaerobic bacteria—bacteria that don't need oxygen to survive. And anaerobic decomposition can lead to an end product containing butyric acid and acetic acid. Those acids cause the putrid smell. And that's what had happened when my compost went wrong. A sour smell is an indication that the soil is low in oxygen and not suitable for growing plants.

Compost is a good model for the processes taking place in a healthy soil system, Linda told me. The same principles apply: you need fungi, bacteria, plenty of oxygen, and the compost needs to mature somewhere that worms, springtails, and other animals have access to it.

"I was visiting a company that made really good compost. And when you stood in the compost bin, you could *hear* the earthworms chewing!" she said.

The World's
Most Important Clumps

When Linda Jolly talked about composting, she used the word "feed"—she *fed* the compost as if the brown mass were full of gaping mouths, slurping and munching on my semi-decomposed leftovers. But what happens to the onion peels and eggshells that wind up in the compost? Food waste is made up of organic material; that is, chemical compounds that contain carbon. Linda explained how fungi and bacteria in compost pick organic molecules apart while simultaneously allowing huge numbers of microorganisms to propagate. The barrel in my backyard is essentially a microbe factory.

"Many of the microorganisms you cultivate in the compost help supply the plants with nutrients when it ends up in the soil," Linda said.

When the food waste has decomposed to the point where it looks like soil, the compost is mature. And when you shovel this compost onto your vegetable beds, it becomes food for bigger organisms such as nematodes, springtails, and other forms of life that—provided everything is as is should be—are living down there. These organisms continue to chew on the leftovers, dividing them into ever smaller pieces, and in turn producing new fungi and bacteria. This organic material then becomes mixed with particles of sand and clay in the stomachs of tiny animals. When one of these animals dies or defecates, these remains will be eaten by other animals thereby passing the baton to the next link in the food chain. Which means good, moldy soil has passed through the digestive systems of thousands of tiny cement mixers and could well be described as a form of recycled poop.

BUT THE TINIEST REMNANTS of these organic molecules are not so easy to break down. One reason is linked to what's perhaps the world's most important clump, or aggregate, as soil scientists call it. The biological activity in soil—such as the physical work of the fungal hyphae, as well as secretions from earthworms and other organisms—binds the clay, silt, and sand to the organic material. This process forms the structures that cause the soil under the grass roots in the forest to become clumpy.[5]

It would be no exaggeration to call them the most important aggregates in the world.[6] They provide optimal conditions for soil hydration, plant growth, and other soil-based life.[7] And they provide optimal conditions for storing carbon in soil in several ways, for instance by sheltering it from decomposition. Inside the aggregates are tiny holes, or micropores. After the microorganisms and soil organisms have done their job with the organic matter, the small carbon-containing molecules that remain can attach themselves to such micropores. It is hard even for microorganisms and their enzymes to penetrate these pores. Protected within these soil aggregates, the small organic molecules can hardly be broken down further. And since these molecules are stuck inside the pores of stable soil aggregates, they don't get washed away when it rains. This means that the carbon can stay in the soil for the long term.

HUMUS IS THE stable end result of the process that begins when I put food waste in my compost bin. When the bacteria, fungi, springtails, nematodes, and all the other organisms have finished eating, and the remaining carbon matter is protected from composting further, you are left with a porous black structure: humus.[8]

The word "humus" and the name of our own genus, *Homo*, originate from the same Proto-Indo-European word $D^h\acute{e}\acute{g}^h\bar{o}m$, which means "soil."[9] The words for human and soil have the same etymological origins in Semitic languages too. But we are connected to humus by more than just a name. One connection is that humus plays an important role in making soils fertile for our food. Another is that humus holds on to carbon so that it doesn't build up in the air. The thin layer of fertile soil on the planet contains two to three times as much carbon as the atmosphere.[10] Humus is, in other words, a stable subterranean reserve of carbon that shouldn't be released into the atmosphere.[11]

However, over the past twelve thousand years, the increase in farmland has disturbed these aggregates and released a lot of carbon from the planet's top layer of soils. The total amount of carbon released from soil during that time is roughly equivalent to eighty years' worth of present-day US emissions.[12]

MY CONVERSATION WITH Linda Jolly was over. I was hoping that I, too, might one day get close enough to soil organisms to hear the earthworms chewing.

Because when I think about soil life, it's primarily earthworms I'm thinking of. Worms are seen, by many, as superstars of the soil. One of the first people to acknowledge earthworms and their underground activities was Charles Darwin, who wrote about them in his formidably titled book *The Formation of Vegetable Mould Through the Action of Worms, With Observation of Their Habits* (popularly called *Worms*). The book—published in 1881, at the end of Darwin's life and over twenty years after he published *On the Origin of Species*—was the result of a forty-year study of earthworms.

Darwin's investigation into earthworms was supposedly started by a curious uncle. A young Darwin was visiting his uncle's garden after returning from his voyage aboard the *Beagle*.[13] This was at a time when earthworms were considered a disease of the soil. The uncle was puzzled about how fragments of crushed brick and ash, which had been strewn over a large area, had over time become buried under a thin layer of soil. He was unable to find any explanation, other than that the perpetrators were the much-hated worms.

Darwin began to collect earthworms, keeping them in pots in his billiards room so that he could get really close to them. He must have spent thousands of hours with his worms. Darwin was interested in whether they had a sense of hearing, so he would get his wife, Emma, to play the piano for them. The worms apparently took no notice of the sounds; however, they did react to the vibrations when he put the pots on top of the piano.[14] He also found that they preferred certain types of food, such as raw fat over raw meat, and concluded that his worms had a passion for eating.[15]

Darwin's book represented a small revolution in how we view worms, and it sold more copies during his lifetime than *On the Origin of Species*.

Darwin was fascinated by how earthworms could make good soil using their tiny food-mixer-like digestive systems. But there were no earthworms under the hedge when I'd buried my underpants. *Were* there any earthworms in the city? If so, where? I began to walk around looking for sticky and lumpy piles of worm feces on the city's lawns. Occasionally the worms themselves would appear—after heavy rain, they lay lifeless and pale on the asphalt.

"Earthworms can have a particularly tough time in the city," said Erik Joner, a senior researcher at the Norwegian Institute of Bioeconomy Research (NIBIO) who was on the phone line. I'd called Joner, who has a PhD in soil microbiology, to ask why I hadn't found any worms in the hole I dug under the hedge.

"In many places I see builders compressing most of the soil when they lay new aggregate. They like to put a thin layer of loose soil on top. But below that, the soil is hard and compact. And worms struggle to get through the soil when it's that hard."

Winter is tough for the worms, according to Joner. To survive the cold months, worms need to burrow deep down into the soil where the frost doesn't penetrate. But when the soil is shallow or compacted, they can't burrow deep enough and easily freeze to death in the winter. This, believed Joner, could be one of the reasons why there are often very few earthworms in compacted urban soil.

"Plant roots can struggle in compacted soil too," Joner said. "Grass may be okay in a shallow soil, as grass has shallow roots. But the root systems of the slightly larger plants and trees, which go deeper, will have trouble. And when the deeper layers are too compact, the soil is less capable of absorbing rainwater. And that leads to flooding."

Over time, a worm can riddle compacted soil with horizontal and vertical passageways. These worm tunnels allow water to drain into the earth and be stored there for use by plants instead of splashing off the surface and creating problems. Eventually the soil loosens up enough for plant roots to penetrate. Worms can also take other organic matter such as leaves into their tunnels, and provide food for the organisms living down there. In other words, it's not just farmers who depend on the efforts of earthworms.

But the earthworm is actually the giant of the subterranean stage. Who are the other millions of organisms living down there?

MY CALL WITH Erik Joner ended and I glanced at my trusty twenty times magnifying loupe. For years it had been all I'd needed. It had shown me the sex organs of lichen and the gripping jaws of black ants. But it wasn't quite enough to see the microbes in soil.

It wasn't the first time I'd wanted to dive into this micro universe. Twenty-five years ago, at some point during the early nineties, I'd had a bowl haircut and a burning desire for two things: a pony and a microscope. A pony for obvious reasons, a microscope because I wanted to see the bacteria and tiny insects living in soil. But despite my intense lobbying, I got neither of them; we lived in an apartment building, and my parents thought I was weird enough already. As an adult, however, I could make my own decisions. I no longer needed the pony. But the life in the soil had been eluding me for a quarter of a century. I needed help. And the protozoa princess could help.

The Protozoa Princess

A tortoiseshell cat jumped onto the table and sniffed the microscope. The image on the computer screen, a pale-yellow moon on a black background, shook violently before settling down again.

"It's curious," laughed the young woman sitting at the microscope. Katelyn Solbakk looks at soil beneath the microscope every day, and has done so for the last ten years. She analyzes samples from farmers and others interested in knowing about the quality of their earth, and is probably one of the country's leading soil experts. Katelyn spoke at breakneck speed in her Ontario, Canada, accent while rapidly twisting the knobs on the side of the microscope. She pushed the cat down from the table.

"I've never set the microscope up like this before, so the cat wants to check what I'm doing."

A camera, balanced on a stack of books and with its lens aimed into the microscope's eyepiece, was connected to the monitor. This setup allowed me to follow both Katelyn and what was happening under her microscope via two separate Skype conversations. The pale-yellow moon on the screen—the microscope's field of view—shook once again.

KATELYN PUT A drop of a light brown fluid on a square piece of glass the size of a matchbox. She then placed another thin square of glass on top and clamped the glass slide under the microscope. The pale circle on the screen was transformed into diffuse gold-colored spots. The cat brushed affectionately against the stack of books, triggering another minor earthquake on the screen. Katelyn pushed the cat from the table again and continued turning the knobs. Amber-colored lumps emerged from the blurred image.

"These are clumps of humus," said Katelyn.

I had learned a little about humus from my conversation with Linda Jolly. But I had never imagined that it would be amber colored. Something about its texture made it look like thick pieces of fabric. Katelyn turned the knobs again, making the image sharper, and the dark-gold spots became large objects with frayed edges. The microscope's field of view began to resemble a photo of an asteroid cloud. Smaller pale-yellow spots and beautiful diamond-like crystals hovered between the larger asteroids, like pieces of debris frozen in space. Katelyn told me that the crystals we were looking at were actually grains of sand.

"Many people who peer into a microscope for the first time feel like they're looking at the moon."

ON THE TABLE in front of Katelyn were three bowls of soil. One was taken from the edge of a parking lot on a housing estate in Katelyn's nearest town. One was from Katelyn's garden. And one was from the nearby forest. The amber-colored asteroid cloud we could see was the soil from Katelyn's garden, dissolved in water, at 100 times magnification.

We zoomed in from 100 times to 400 times magnification, and the asteroid cloud began to come to life. Something moved, but when I tried fixing my eyes on it, it vanished. Then I noticed something else, in the middle of the screen, that looked like a plastic bag full of water.

At first it appeared to be standing still, but after staring intently at it for a while I could see that something inside the plastic bag was moving.

"It's an amoeba," said Katelyn. "It eats bacteria and lives inside a little transparent house."

I took a closer look. The transparent bag was covered with tiny transparent shells, like a fish. And in the middle of the bag I could just make out a circular hole.

"When it's going to move, it pokes a tiny bit of its soft body out of the hole and moves about like a snail, with its shell on its back."

Something tiny, perhaps a hundred times smaller than the amoeba's length, approached the amoeba with dazed, spiral-shaped movements. And this *something* was clearly in trouble; it was moving about like a mouse with its tail stuck in a mousetrap. Katelyn explained that it was a ciliate that had become trapped in a piece of debris its own size. Now the ciliate was desperately trying to free itself, spinning like crazy one way before changing direction.

"This little thing is about two micrometers long, in other words, one thousandth of a millimeter. But it's quite obvious that it has a problem that it's attempting to solve."

I studied the animal writhing in despair in its own dance of death. I'm not quite sure what I was expecting from an organism consisting of a single cell. But I wasn't expecting to feel sympathy.

Ciliates belong to the phylum of protozoa, the single-celled organisms that gave the protozoa princess her name. The Greek word *protozoa* means "primitive animal." But that doesn't mean these organisms are primitive. They just found out early on how they could conquer the world with their single cell. A single cell that hunts for prey, mates, and lives in a way that is anything but passive.

The ciliate moved as though it had a will of its own. Katelyn told me how its movements are based on chemical signals. "This is of course essentially how the human brain functions too. What's the difference between their desire to live and our own? Looking under a microscope can make you quite philosophical."

While studying at university, Katelyn was the only person who analyzed soil samples under a microscope. She was so excited about the protozoa she saw that she called people over to have look. That's how she got her nickname. "The protozoa blew my mind. I never got over it!" laughed the protozoa princess.

The little ciliates on the screen rested for a moment. I could see a few tiny hairs sticking out from one end of the transparent creature, but its long tail, which must have been what snagged on the small, light-yellow piece of debris, was invisible. Another ciliate moved across the screen, as determined as a clear thought.

I asked Katelyn why there are so many protozoa in garden soil. What were they doing there? She explained that there is a circular economy going on in the soil: nutrients are passed around in a cycle between plants, microorganisms, and soil. And the protozoa are an essential part of this cycle.

In the nutrient cycle, bacteria break down dead organic matter, such as dead leaves, that contains a lot of nitrogen. But bacteria have a real thirst for nitrogen. They absorb it and conserve it. And if there's nothing to eat the bacteria, a lot of nitrogen will be trapped in the large number of bacteria living in the soil.

Protozoa don't need that much nitrogen. What they are most interested in is the sugar and other substances the bacteria contain. When protozoa graze on bacteria, they release a lot of nitrogen back into the soil, making it available to plants.

Thanks to Katelyn's work, one of these billions of soil organisms—one of the anonymous nuts and bolts that keeps the cycle going—had gained a personality as it spiraled around on the screen.

KATELYN TWISTED THE KNOBS AGAIN, leaving the ciliate to its dance of death, and scanned more of the asteroid cloud.

"Oh!" she suddenly exclaimed, focusing on what looked like a flower on a stalk except for its puckered lips instead of petals. Rotifers, she explained, were so named because their tiny mouth parts can look like rotating brushes, which they use to sweep up and eat bacteria, plant remains, and other free-flowing dirt.

WE SCANNED THE IMAGE a bit more. Something long, thin, and transparent twisted frantically. It was much larger than the rotifer and looked like a curly pubic hair that had come to life.

"It's a nematode!"

Nematodes live in the soil but are actually freshwater animals. Like nearly all other tiny, soil-dwelling organisms, they swim in the membrane of water surrounding the particles in the soil. To them, soil is a network of tiny lakes.

"Nematodes have a bad reputation," said Katelyn. "Some of them, the root-knot nematodes, are indeed parasites that live on plant roots. But the vast majority of nematode species have a positive effect on the soil and plants."

Katelyn explained that some nematodes graze on fungi and plant remains. Others pierce tiny animals with their pointed mouths, injecting a bacterium to kill the animal before sucking out its insides. By doing this, nematodes help to keep the numbers of larger, and for humans more troublesome, insects—such as fly larvae—under control.[16]

"If you have a variety of organisms in the soil, it will regulate itself. Nematodes are an example: it's okay for your soil to contain bad nematodes, as long as it contains lots of other species that eat nematodes. Nematodes can also kill insects and animals we don't want," says Katelyn.

SUDDENLY A MASSIVE CLAW entered the field of view before disappearing again. Katelyn stopped looking at the nematode, and after a short hunt, aided by the knobs on the side of the microscope, found the claw—which turned out to be clinging onto a long, hairy limb, a bit like a crab leg. I had to fight my revulsion. I've never been keen on tiny hairy-legged bugs, especially when coming across them unexpectedly. This leg was many times thicker than the nematode, and it was huge compared to the ciliates.

Katelyn zoomed out. The hairy leg seemed to be attached to an absurd-looking creature. Its head, which was similar in shape to a rabbit's but with antennae instead of ears, was attached to a hairy shrimp-like body with six legs. These legs were kicking in an attempt to free the creature from the air bubble it was trapped in. Something inside it pulsated, like a heart beating furiously. It was like watching a rabbit caught in a trap.

"A springtail," said Katelyn. But where a rabbit would have front teeth, I could see a little beak, which, as Katelyn explained, is used for grazing on dead organic matter. You can find springtails at the intersection of living and dead matter in the soil.[17]

Springtails also have one of the animal kingdom's more peculiar mating rituals: the female will lift the male and twirl him around in some kind of partner dance until he releases his sperm onto the ground. She will then pick up the drop of sperm and complete the fertilization process by shoveling it inside herself.[18]

But the springtail under our microscope was in a different mood entirely. Its articulated tail was folded under its body like a tiny shrimp, a coiled spring that can be released if attacked. I had seen springtails before on the late-winter snow, and they were tiny black specks that leapt all over the place. To see one of these creatures up close, with its insect-blood pumping through its brown-red body, was almost a shock.

WHETHER THE CREATURE had hairy legs or not, the nine-year-old within me was already shrieking with excitement. The space Katelyn was exploring was full of creeping, crawling, quivering life. The amber-colored asteroid cloud under the microscope was beautiful and strange. Perhaps the moon wasn't such a bad comparison.

We then noticed long streaks that crossed the field of view like thick cables. At first I thought they were stray hairs, but they seemed too stiff, and some of them had branches.

"Hair is much, much thicker," said Katelyn. "No, this is fungal hyphae."

What we recognize as mushrooms, the hat-wearing stalks that we pick and eat, are just a small part of a fungus. They are its "genitals," which poke up from the ground and spread the fungus's spores. The rest—the main part of the fungus—is made up of hyphae, which are long cells that penetrate the soil and weave thick networks. A hypha can be several miles long. So, the fungus's power and influence are actually down in the soil.

A honey fungus (*Armillaria ostoyae*) in Oregon is the largest-known single organism in the world.[19] This decomposing fungus covers an area of almost four square miles and could weigh as much as 35,000 tons. That's roughly the equivalent weight of an eighty-seven-mile traffic jam.

"FUNGI OFTEN THRIVE BEST in soil that's left fairly undisturbed," Katelyn said. Fungal hyphae, the long threads that make up the fungi's underground body, are easily damaged by digging.[20] You can compare it to an excavator digging randomly around the main square of a big city. It would sever quite a few power and internet cables.

But when the hyphae are left to grow in peace, they help give the soil some important qualities. Fungal hyphae and other microorganisms virtually glue minerals, sand, and humus to the stable soil aggregates that help protect the carbon in the soil. The hyphae in some fungal species also break down very slowly, which helps to capture the carbon. In some ecosystems, the combined plant roots and fungi account for between 50 and 70 percent of the soil's carbon storage.[21]

Something in the Air

Katelyn pointed at the large, amber-colored lumps with frayed edges swarming on the screen.

"These are pieces of soil aggregate, full of different-sized holes and pores. Between one- and two-thirds of the soil's volume is made up of these pores, which contain either air or water."

I remembered what the soil in the forest had felt like when I buried the cotton underpants. It had been like holding a fluffy-yet-sticky piece of chocolate cake in my hand. Soft and rather light. I could squeeze it together without it crumbling, perhaps because its tiny holes had collapsed like the air bubbles in a piece of sponge cake when it's forced into a tiny box.

The soil in the city, under the hedge, had disintegrated between my fingers like an overcooked brownie.

WHEN IT RAINS, the pores in the soil fill up with water. Surface tension keeps the microscopic water droplets trapped in the smallest pores, making them hard for plant roots to get at. The larger pores fill with water too, but the surface tension does not retain this water as firmly, which makes it easier for the plant roots to obtain. Good soil contains a lot of pores, at different sizes, giving it a huge capacity for storing water. On a hot day, a large urban tree can absorb more than 100 gallons of water. So, its root system must be connected somehow to thousands of gallons of water if the tree is to remain hydrated through a couple of weeks of dry weather.

It is in this world of air- and water-filled pores, tunnels, and caves, that soil life exists. Just as fish and other animals hide in the nooks and crannies of a coral reef, the tiny soil organisms shelter in the tiny water-filled pores of the soil structure.

When heavy vehicles drive across soil, the water-filled channels and the large pores that aerate the soil get compacted, like the cake squashed into the tiny box. It forces the air out of the large soil pores, and leaves less air for the animals and fungi living down there.

What happens to the mites, nematodes, and long fungal threads in soil when it is compacted, dug up, and moved? I asked Katelyn about how robust these underground ecosystems really are.

"The fungi and many of the animals are highly vulnerable to what we do to the soil," she replied. "In compacted soil, which city soil usually is, there are often more bacteria, and fewer fungi and animals."

According to Katelyn, fungi and bacteria take a more or less equal role in breaking down organic waste in soil.

But fungus doesn't thrive in compacted soil that has been disturbed. Instead of a balance between decomposers, bacteria predominate in disturbed soil. This means that many of the other soil-dwelling animals—such as springtails, which eat fungal threads—disappear.

Often the animals high up in the food chain, such as predatory nematodes, will vanish first if the soil becomes too inhospitable. When that happens, the balance between animals, fungi, and bacteria can change in ways that make the soil less optimal for the animals and plant roots down there. When the system doesn't work, the soil's available nutrients—the building blocks needed for growth and survival—can run out.

"There will then be a collapse. Very little will grow."

COMPACTED SOIL LACKS the pores and passageways that roots need to grow in. Even if roots do manage to penetrate hard soil, there's very little water in it for them.

The plants that thrive in soil that is compacted and low in organic matter are those with deep and powerful root systems. One species that sprang to mind was mugwort, a gray-green plant with hairy, lobed leaves that is also the scourge of many allergy sufferers. Another was dooryard dock, also known as northern dock, a tall, tough-stemmed plant with small light-green flowers that turn into brown-colored seeds in the fall—and these seeds and flowers are like confetti. Anyone who has tried pulling up mugwort or dooryard dock will know that they have long, powerful roots, and both like to grow where other plants can't—straight from a seam joining asphalt and brickwork, for example.

I told Katelyn about the biscuity soil I had dug up beneath the hedge. There was no sign of life in it, not one single earthworm, and no trace of the lumpy structure that the forest soil had. Nor were there any different-colored layers.

"I don't know if what you're describing is a total collapse. But it doesn't sound good. Earthworms can tell us about the conditions farther down the food chain. If you don't see any worms, it can be a bad sign."

WE WERE FINISHED with the garden soil for now. Katelyn measured out a tablespoon from the bowl of urban soil and emptied it into a small jar of water that she gave a good shake. Using a pipette, she then dripped a light-brown drop onto a new slide, placed a small square of glass on top, and swapped it with the glass slide currently under the microscope.

I was not prepared for how different the urban soil and garden soil would be. In the garden soil I'd seen large amber-colored clumps of humus, sand crystals, and spiraling creatures darting purposefully between them under the microscope. Now the screen was filled by a collection of tiny evenly spread particles. There were no large amber-colored meteors, just a few pale-yellow lumps that looked like they were about to fall apart. Instead of large, thick lumps of humus and organic matter, the screen was filled with little bits of debris and transparent diamond-shaped crystals. Sand.

If the previous sample had been a quiet Sunday afternoon at the park, the image on the monitor now looked more like a community that had been blown to smithereens.

"That's almost exactly what's happened," said Katelyn.

This soil sample had come from a parking lot. Neither of us knew exactly what had happened to the soil there, but it had certainly been dug up and moved around. It may even have been transported from somewhere else.

Katelyn zoomed in closer on the sample. There were no amoebas lazing around in their shells, or protozoa darting back and forth. But after some hunting around, Katelyn found a tiny thing spinning through the particles. "It's a protozoa, a small flagellate typical of those I find even in the most ruined soil. It's the first life-form to return when soil has been disturbed."

What we could see was a community that was falling apart. Or was it a community in the process of rebuilding?

"Ah, great!" exclaimed Katelyn suddenly. The dark cables, which I knew were fungal hyphae, had entered the picture. "There's actually a decent amount of fungal hyphae here. I wasn't expecting that."

We followed one of the dark threads, which was clinging to a long, thick object that darkened the whole screen. It was hard to comprehend that this huge object was actually a hair-thin plant root, and it was enveloped by masses of fungal hyphae.

"This could be why there's so much fungal hyphae," said Katelyn. "Plants help soil communities rebuild." Even the weeds, although unwanted, play a role in kickstarting soil processes again. Plants help the soil heal.

She said that a weed's solid roots can penetrate the soil, enabling air and water to reach the microbiota. They also provide food for all the tiny organisms living down here. The soil we were looking at from the parking lot had been dug up and its structure destroyed. But the soil had started the process of regenerating.

"There's another flagellate! Thing's aren't as bad as I'd expected. It's much better than some of the farm soil I've looked at."

According to Katelyn, the most damaging thing for soil is perhaps repeated digging, or plowing. "People often ask me what they can do to improve the soil in their gardens. Well, the most important thing you can do is leave it alone." [22] Fungi, bacteria, and microscopic animals can handle being dug up and exposed to light and air once or twice—but they don't stand a chance if the soil is plowed repeatedly. Plowing is a powerful weapon against soil microorganisms.

"I see hardly any life in soil that's been plowed, sprayed, and amended with artificial fertilizer for years. Instead of allowing the soil to do its job, we try to do it ourselves and end up destabilizing the system. We end up suppressing the very thing that actually helps us," she said.

"I don't like using terms like 'good' and 'bad' soil. It depends on your perspective. But some soil types are more beneficial for humans, and the ecosystem's other inhabitants, than other soil types are."

"SURELY IT'S POSSIBLE to grow plants in soil that comes in plastic bags and is free of microorganisms?" I asked. I'd remembered something Erik Joner from NIBIO had said: "It's possible to grow plants in the sterile conditions of a greenhouse. You can even use hydroponics, which only requires water. But you then have to do a lot of the work yourself. You have to give the plants nutrients and make sure the conditions are right. You get no helping hand from the microorganisms."

Katelyn swapped the glass slide under the microscope with one containing dissolved forest soil. Once again, the screen was filled with large, dark, amber-colored clumps held together by fungal hyphae. But these cables of fungal hyphae were even more tightly clustered than those we saw in the garden soil sample. The screen was covered by a tangled mass of light and dark threads, which branched out from one asteroid to the next.

In one corner, the transparent curl of a nematode squirmed in what looked like a futile attempt to tie itself in a knot. There were so many amoebas on the screen that it looked like an ocean full of diaphanous shopping bags. It took me a while to distinguish these vibrant, living shapes from the clumps of humus. But suddenly I saw just how numerous and varied the amoebas were; from long, articulated, glassy strands containing what looked like droplets of green oil to tiny pearls of transparent amber.

In another corner, something dark and grotesque appeared: a severed foot with dinosaur-like claws and armored segments on the leg. It looked like the whole Star Wars universe had been put into a food processor. Perhaps the foot belonged to a moss mite?

The moss mite, a relatively large micro-beast, wanders around on four pairs of jointed legs while chewing on mushrooms and rotting plant parts, like the soil-world equivalent of a cow. Moss mites crawl through the soil's network of tunnels, which are full of microorganisms, and spread these fairly immobile microbes elsewhere. Sometimes moss mites will hitch a ride themselves: loaded with fungi and bacteria, they'll attach themselves to an insect—a bumblebee, for example—and fly long distances, taking the fungi and bacteria to new places with fresh soil to colonize.[23]

TOWARD THE END of our conversation, Katelyn talked about how difficult it is for her to form a clear picture of the lives playing out in the microscopic world of amoebas and ciliates.

"Even what we see under the microscope is a false image of what's happening," said Katelyn. First of all, she explains, the soil and its organisms look different underground, in the dark, than they do when dissolved in water and squeezed between two thin sheets of glass. What we see under a microscope are soil aggregates that have been torn apart and organisms that have been thrown into chaos. The other thing is that the extremely strong light from the microscope makes the cells look flatter and more transparent. The amoebas, protozoa, and nematodes are actually far more three-dimensional than the glassy contours visible on the screen. Viewing soil life under a microscope is like looking at X-rays of people and thinking that you understand what a human being is.

The Superorganism

Viewing soil under a microscope was wildly fascinating. But it still wasn't enough to get to the bottom of what was really happening in that soil. I needed to delve deeper.

"Follow the money," say journalists who want to reveal how the world is really connected in order to shed light on the powerful and their murky secrets. Follow the money. And in the world of soil, carbon—organic matter—is the currency.

Carbon can enter the soil in the form of dead animals, or dead plants as it does via the compost soil I put on my vegetable beds. When the large carbon molecules in dead matter are broken down by fungi and bacteria, that carbon can be stored in the soil in the form of humus, the small residual molecules that are protected from decomposing further.[24]

But other processes, especially those associated with intensive farming methods, can lead to a reduction in the amount of carbon that binds to the soil.[25] Farming vegetated areas that haven't changed for generations, such as forests or meadows, can cause the loss of up to half of their carbon reserves. The organic compounds are broken down, and the carbon molecules within them released into the air as carbon dioxide. Complex interactions between many factors govern the extent to which this release of carbon into the atmosphere occurs.[26]

Carbon, the currency of soil, can escape from the soil, where it does good, into the atmosphere, where it can be harmful in large quantities. Because Earth's crust contains so much carbon, small changes to the soil's carbon levels can have a major effect on the carbon dioxide in the atmosphere, which in turn can have a major effect on the global climate.[27]

But there is another process, an ancient underground collaboration, where carbon goes the opposite way—as in, from the air into the soil.[28]

It was this collaboration I wanted to know more about when I rang Klaus Høiland a few weeks later. He was a professor of mycology, or fungal science, when I studied biology. Now a professor emeritus with an office at the University of Oslo, Klaus was a legend among students. His lectures featured a small yellow cartoon bear that wore blue shorts, a superhero called Snutetute ("Snout Spout"), who traveled back and forth in evolutionary time and told cheesy jokes. Klaus himself was well known for his good humor and ceaseless enthusiasm for anything to do with fungi. I was glad to have an excuse to call him.

"FUNGI, BACTERIA, AND ROOTS work together like one great superorganism!" said Klaus on the other end of the phone line, no doubt waving his free arm in the air.

Some fungi, collectively known as mycorrhizae, survive by collaborating with plants and trees.[29] These long, thin fungal hyphae that live underground are far better at finding nourishment and absorbing water from the soil's tiny pores than the (relatively) short and thick plant roots. The fungi deliver any excess water and nutrients they have to the plant roots. In some ecosystems, Klaus explained, especially ones that are low on nutrients in the first place, the plants can get up to 90 percent of the nutrition they need from these fungal roots. In return, the fungi gain access to the plant's sugar.

Carbon dioxide in the air is converted into sugar in a plant's leaves via photosynthesis. But a plant won't use all the sugar itself. Between 10 and 20 percent (some researchers think it could be up to 40 percent!) of all the sugar that a plant produces seeps out as a sticky, carbohydrate-rich mass in and around its roots, where a community of microorganisms awaits. This ancient transaction—which involves fungi mostly, but also bacteria—essentially takes carbon out of the atmosphere and returns it to the soil.

A teaspoon of soil can be interwoven by almost a mile of different fungal hyphae! I pictured an intricate network of wires and cables just below the soil with fibers running in all directions. The trees use the network of fungal hyphae to communicate, warning each other of attack, for example. A tree infested by insects will send stress molecules into the ground. The signal will be detected by the fungal network connected to the roots. Thus, the trees connected to this underground network will interpret the signals and produce substances to protect themselves against the insects—before they too are attacked.[30]

This superorganism works best if the fungus, the plant roots, and the other microorganisms in the soil are all happy. Soil that does not have a supply of dead material or isn't covered by plants that can pump it full of sugar simply won't get any food.

"A lot of urban trees struggle," said Klaus. "Especially when there's a drought. And the lack of underground fungi is one of the reasons for this struggle."

I WOUND UP my conversation with Klaus, and went out. It was October, and one of those warm fall days when you almost forget that frost is just around the corner. Beams of warm light streaked between the slender birch trunks, and the air felt easy to breathe. In the botanical garden I found a patch of grass between the trees that was covered with yellow-brown birch leaves, and I lay down.

When you lie on the grass like this, and there's nothing else to look at, you might find yourself studying the few remaining leaves hanging from the long, drooping birch twigs as they are caught by short gusts of wind. When one of the leaves finally lets go, you might then follow its random descent as it eagerly makes its first and final journey before settling among the other leaves on the ground. If you've got plenty of time, you might find animal tracks among the piles of fallen leaves. Passing dogs, people, or other animals may have flipped over some of the leaves, revealing their darker, half-rotten underside; and the leaves that have lain at the bottom will be a step ahead in the process of becoming soil.

In some places, you might find a leaf poking up vertically, neatly rolled like a pixie hat with its pointed end down. One such leaf caught my eye.

I'd never seen a birch leaf fall to earth shaped like a pixie hat before. I got up and walked over for a closer look. It was actually more like a loosely rolled cigarette, and was in fact on its way into the ground. A tower of something gray and lumpy, and an inch or so high, revealed what was to blame. What clever worms, I thought, impressed by their idea of rolling the leaf before dragging it into the ground.

These pixie hats were a reminder of all the unseen things going on around me. Something had been here. Something *was* here. To understand what was going on under the ground I would have to interpret the visible signs on top of it, like the traces found at an archeological dig. If only the soil were transparent so that I could *see* the earthworms! Not to mention the fungal threads which I imagined looked like a layer of cotton candy just below the surface.

ON THE GRASS NEARBY lay a round, reddish object about the size of a Ping-Pong ball. When I reached down to pick it up, I discovered it felt leathery. To my surprise, the ball was also stuck to the ground. When I pulled, it made a light, crisp-sounding crack as it broke off in my hand. I turned the object over and saw that its underside had a spongy, pale-yellow surface. It was no ball. It was the cap of an orange birch bolete, *Leccinum versipelle*.

Leccinum are a genus of fungi that *always* cooperate with trees. The orange birch bolete, however, is one of the more picky ones. Its hyphae will attach to birch tree roots only. So, the mushroom I had in my hand had drawn most of its energy from the surrounding birch trees. The orange-red ball was aboveground evidence that fungal hyphae stretched beneath my feet. A network of fungal threads that was perhaps connecting all the great, old birch trees that surrounded the lawn.

Could the mushrooms on the surface tell me anything about the city's underground history?

THE MAP ON the screen in front of me was covered in red and green dots, all clustered like shotgun holes in a target. Between each cluster was just the occasional red or green dot. It was a map of the city center, and every dot was a mushroom.

I was sitting in front of my laptop in my kitchen. The map in front of me was Artskart, an online map service similar to iNaturalist, that allows you to search for

Norwegian species and genera, then shows where in the country they have been found. People who have found mushrooms, animals, birds, and other organisms can register their findings on the map.

I searched for three of the fungal groups that live exclusively by cooperating with trees: the *Lactarius* and *Russula* genera, and the bolete family. *Lactarius* are gilled mushrooms characterized by an apple-like, mealy flesh. And when they are broken apart, a milky fluid trickles out— which is why they are commonly known as milk caps. *Russula* are also gilled mushrooms and have an apple-like flesh, but they do not contain the same milky fluid. The boletes are not gilled; instead, they have a layer of narrow tubes beneath the cap, just like the orange birch bolete.

The mushroom finds displayed on the maps on my screen, like the mushroom maps of other big cities, looked like tiny mushroom islands in a vast ocean. These islands were located mainly within the city's parks and green areas. And the sea—that is, the rest of the city—was somewhat dot free.

I knew that the botanical garden had been established in the mid-nineteenth century. Its open spaces have never been built on or paved, and the soil's fungi have been left more or less untouched. And judging by the mass of colored dots on the map showing the botanical garden, my orange birch bolete had company.

And not just in the gardens. The city's graveyards were especially packed with such cooperative fungi. Old images and aerial photos showed that, just like the botanical garden, these graveyards have lain mostly undisturbed for many years. When the first aerial photos of Oslo were taken in the 1930s, the trees now standing in the grave-yards were already there.

It might be that there are simply fewer places for fungi to poke their caps up through the asphalted streets. Perhaps they like the variety of trees in the parks. Or perhaps the mycologists, mushroom experts, recording their finds were just attracted to cemeteries? I wasn't exactly doing a controlled scientific experiment, of course. But I believe that what attracted these fungi to the old parks and cemeteries was primarily the soil's age.

The dead and buried are difficult to move. Graveyards are rarely dug up to create space for a growing city. So, a city's graveyards and parks are islands of continuity in a sea of constant change. And just like the deceased, cooperative fungi like peace and quiet.

The Tale of the Underpants

I had been tempted to find out more about soil life. The idea came out of the blue one day while I was out driving and the local winner of the annual Soil Your Undies campaign was announced on the radio. Soil Your Undies is where farmers all over the country bury a new pair of white cotton underpants on their land for a couple of months. All the underwear had just been dug up, and the winner—the farmer with the dirtiest, most holey underpants—had been declared.

This was not an outbreak of collective madness. Burying a piece of cotton for a few weeks is one way of measuring how healthy your soil is. The more brown and decomposed your underpants become, the better the soil life. What actually breaks the cotton down is, of course, the team of bugs, fungi, and bacteria.

I had buried nine pairs of underpants: three in the forest, three in my compost bin, and three in the patch of urban soil under the hedge, where they would lie for a few months until I dug them up again. The result of the microorganisms' work would indicate how much life was down there. As I patted down the last bit of soil, I felt like I would soon be literally unearthing the truth about soil.

I waited until November. It had still been summer when I buried the underpants; the trees had been a dusty green, the colors soft and watercolor-like. By November the day looked like it had been drawn by a recently sharpened pencil. The bright light was softened by high clouds, and the black and bare twigs on the trees rattled in the wind. I was on my way to check the state of the underpants under the hedge in town. Three months in the soil should be enough, I thought, even after a few weeks of cool fall weather.

The cotton underpants I'd buried in the compost bin had disappeared in a few weeks. Each time I checked on them they looked more like the surrounding brown compost, until they finally vanished completely. All that remained were the elastic waistband and the polyester label, both of which would inevitably show up in the finished compost along with other slow decomposers like avocado stones and fruit stickers.

The underpants buried in the small forest beside the subway stop had been difficult to find. The ground had been covered in brown leaves, and it took a long time to locate the popsicle sticks I'd used as markers. There, too, the underpants had settled so comfortably into the soil they had almost become a part of it. The fabric was tattered and had turned a gray-brown color. The only things still resembling what I'd initially buried were the elastic waistband and shiny white label. I was so excited by how much had decayed that I gave a little cheer.

Finally, it was time to bring the life under the small inner-city hedge into the light.

I rounded the corner where I had buried my underpants. Two kids wearing snowsuits watched me from a brand-new, bright yellow jungle gym. The space that had previously contained a sad-looking hedge in a little square of soil was now covered by artificial grass tightly abutting the asphalt on all sides.

Where were the underpants? Where was the soil they had been buried in? What about the fungi, the bacteria, and the nematodes, rotifers, and springtails that might or might not live down there? I stood and looked around for a moment, as though the hedge had just gone for a walk and would soon return, or as though I might find a note from the landscapers informing me about the missing underwear.

Could this be the first time in the history of our species that many of us walk around without ever being in contact with soil? Without touching it, without digging it, almost without seeing it?

"THEY DON'T JUST make our ecosystems work; microorganisms can also have a direct effect on our health." Katelyn Solbakk was on the phone. She had called to say that there are many indications that soil makes us healthier.

Researchers have found that exposure to the soil-dwelling bacteria *Mycobacterium vaccae*, a decomposer, can prevent stress-induced inflammatory conditions in lab mice and also change their production of serotonin.[31] Low levels of serotonin are associated with a number of disorders, such as anxiety and depression. Mice that were exposed to soil bacteria simply became less ill from stress than mice that didn't receive such treatment.

The researchers explain the findings through what they call the "old friends" hypothesis: important processes in the immune system are driven by signals from microbiota, including organisms to which humans are exposed in their environment. They believe that the lack of contact with these microorganisms, in modern urban societies, is one of the reasons for the ongoing epidemic of inflammatory conditions. The researchers propose that humans should be reintroduced to their "old friends," the immunoregulatory microorganisms.

Finnish researchers followed up on these findings by transplanting segments of forest vegetation and soil to day-care centers in urban areas. After the children had played in the heather and forest soil for a month, researchers tested the children's bacterial flora. They found that the bacterial communities on the children's skin and other areas had changed in a way that could improve their immunoregulatory systems.[32]

Katelyn believed that soil also helps lift your spirits. "We normally take a bucket of soil to our events," she said, "and encourage people to put their head in the bucket if they're feeling gloomy."

I ENDED MY CONVERSATION with Katelyn and went outside into a clear November day. In the backyard, the earthworms were clearing up after the summer. The mountain ash's yellow leaves covered the small patch of lawn. I never managed to catch any worms in the act of drawing leaves into the soil, but I could see the tiny leaf cones poking up from the ground. Each time I walked past, they had been tugged a notch farther down, like a volley of arrow tips slowly piercing the earth. Soon the earthworms below would plug the entrance to their tunnel with a leaf, and then curl up and go to sleep in anticipation of spring.

I thought about how urban soil is a human-made patchwork, where each patch tells a different story which in turn makes it a map of a city's history. Even urban soil isn't dead matter; it is a living system. And like all life, it has an enormously powerful life force.

Erik Joner had said that urban soil that has lain undisturbed for a couple of years can be fairly sustainable. Even if soil is initially bad, it doesn't take long for it to improve if the plant roots, earthworms, and fungal hyphae are left alone. Had it not been dug up, even the small patch of soil under the hedge would have been reclaimed by nature. Bumblebees would have landed on the hedge's flowers and brought tiny bugs, such as moss mites, loaded with fungi and bacteria. Had the soil been fully vegetated, the plant cover would have helped the soil maintain a more consistent temperature and humidity. The plants would also have sent sugar into their roots, which would have seeped into the soil and provided food for the life growing down there.

Linda Jolly was also optimistic about urban soil's ability to bounce back. She had succeeded in getting the soil's microbiota up and running in places that would seem impossible. "It's important to keep soil well vegetated, to give the animals a roof over their heads," she said. "The plants provide food for the microbiota down there. Don't do any unnecessary digging. Feed the soil compost, which releases nutrients slowly and continuously, and also enriches the microscopic life with billions of tiny living creatures. Plant lots of different species to give the bugs and microorganisms, both above and below ground, some variation in their diet."

And a few days after we had looked down the microscope together, I received a text message from Katelyn: "Found water bear in driveway. Life is resilient!"

(WATER BEARS, OR TARDIGRADES, are a group of slightly cute micro-animals that, as their name implies, look a bit like bears. They can survive in most environments, but the fact that they can live among mosses growing in cracks in the asphalt is still a pleasant surprise.)

Marvels of
the Darkness

T HE CITY FELT DIFFERENT on this October night. The fog and darkness sharpened the senses. And as I stood on a footbridge linking Oslo's east and west sides, I could feel the damp air clinging to the walls of my nostrils. Brick buildings that flanked the river vanished into the fog, while soft white cones of light seemed to rise from the ground below the streetlamps. The thundering sound of the river almost drowned out the traffic noise from the bridge a short distance away. It had been raining for weeks. There was almost no one around, except for two heavily made-up teenaged girls who passed me on the bridge. A frost was on the way.

As I looked down at the smooth, black river running below, a tall man in his fifties came striding toward the bridge from the east. It was Kjell Isaksen. Kjell is a biologist and one of the country's leading experts on Norwegian bats. When he spotted me, the chiseled face beneath his hood broke into a friendly smile. An old bum bag slung round his waist suggested that he is perhaps an avid cross-country skier during the winter.

After briefly discussing the plan for the night, we walked toward the footpath that runs alongside the river. But instead of continuing along the illuminated path, Kjell did a sudden ninety-degree turn, tramped across the withered grass, and disappeared into a wooded embankment that slopes down to the river. Leaving the safety of the footpath felt slightly uncomfortable, and I soon regretted wearing my light jogging shoes when I slipped on a pile of leaves and skidded the last little bit to the riverbank on my ass. I brushed the leaves and soil from my trousers while Kjell stood and listened. Above our heads, a dog and its owner crossed the bridge that looked almost as if it had been slung over the river like a string of lights. The walkers didn't see us, confined as they were in their tunnel of light. But my eyes were now adjusting to the dark, and it was like discovering a secret room in the middle of the city.

In the cold and clammy darkness, Kjell reached into his bum bag and fished out a small black rectangular box called a Pettersson ultrasound detector D240x. He connected some headphones to the box and gave them to me.

The dominant noise came from the river. But the moment I put the headphones on, I was able to hear other things too. There was a gentle hissing sound, like the rustling of a swan's feathers as it comes in for a landing, which came and went at random interludes. I also heard some kind of clicking sound, like raindrops hitting a soft surface at rapid intervals. I took the headphones off again. Silence.

The Pettersson D240x is a bat detector, a device that converts the bat's high-frequency sounds to a range that is audible to humans. We're essentially deaf to the noises bats make. The audio waves they emit are too dense and at too high a frequency for human hearing. The sensory organ of the human ear, the spiral organ, can (with individual variations) hear up to about twenty kilohertz (kHz), twenty thousand oscillations per second. Most insectivorous bats make calls with frequencies between twenty and sixty kHz.[1]

The black box attached to my ears was a thing called a time-expansion detector. It stretches time, converting the bat sounds to a range audible to humans by reducing the frequency of the audio waves, a bit like a tightly coiled spring being pulled in both directions. A three-second recording, for example, will be stretched to thirty seconds, thus making the frequency one-tenth of the original. So, a bat call at fifty kHz will become five kHz—well within the range of human hearing.

With the headphones on, I should be able to hear like a bat.

I put the headphones back over my ears. The hissing noise was still there, and the clicking sounds were getting louder. I looked around in confusion. Knowing that something was there, that I couldn't see or hear with my normal senses, was quite uncomfortable.

A car passed by on the road a bit farther along the river valley. It stopped for a pedestrian, and just as it braked, the hissing noise returned. But when I removed the headphones, I couldn't hear a thing.

"It might be the brakes," said Kjell. "Perhaps they make high-frequency sounds we don't normally hear."

Whoever was doing the clicking, they didn't make themselves known. But I no longer felt uncomfortable; it was exciting, like having a superpower such as X-ray vision or telepathy.

THE FIRST TIME I can remember hearing about bats, I was about ten years old. I was going on a fall-weekend cabin trip with a friend and her family. The cabin overlooked a huge lake and had no electricity or running water. While we were driving there, my friend's big sister told us there were bats in the surrounding forest and that you could see them hunting at dusk, dancing across the surface of the lake and around the trees by the shore.

That night, as we lay in our bunks talking, my friend's sister went on to say how bats could smell you in the dark, that they crawled over your body until they found an exposed patch of skin and bit down hard, so hard it was impossible to get them off. And they would just hang there sucking your blood, like a tick the size of a tennis ball, before flying away at daybreak.

There were no nocturnal visits to the outdoor toilet for me that night. Instead I lay awake, terrified and desperate for a pee, while looking forward to going home to the city.

I LATER FOUND OUT that you should never believe the scary stories big sisters tell. But I couldn't shake the image of the tennis-ball-sized bloodsucker, and that image developed into a general aversion to bats. For many years a bat has been hunting around the streetlamp in my backyard. I have sat countless times on the veranda on fall nights, watching it move like a piece of flickering darkness. Following the outbreak of COVID-19, which some suspect may have originated from bats, I'd been hoping it would stay well away from the backyard.

But equipped with my new audio superpower, I wanted to see if my relationship with bats could be improved. After all, there aren't many wild mammals that I can watch hunting from a ringside seat on my own veranda. Because bats are right here among us. A friend told me that she used to play badminton in the park and that a bat would come around when it began to get dark. She believed that the bat was interested in the feathered shuttlecock. Once they've first noticed bats, the city dwellers I know see them fairly often, flying erratically around the streetlamps after dark.

THE BATS WE were listening to that October night were from the order Chiroptera. Or hand-wing, if you translate the word directly from Greek (*cheir* means "hand" and *pteron* means "wing"). The name reflects how bats and humans share some anatomical traits: bat wings have structures that are analogous with human arms and hands. If you hold a bat wing up to the light, you'll see that the wing membrane is primarily supported by four narrow, bony fingers, like the ribs on a black umbrella.

Chiroptera, or hand-wing, is an appropriate name in other ways too: when you're going to create a bat's shadow on the wall, you just hook your thumbs and spread your fingers out on either side.

Leonardo da Vinci had good reasons to use the bat wing, rather than the bird wing, as a model for his flying machines. Because where "wing-fingers" are largely joined, bats can move the fingers in each wing individually, which enables them to fly in a way that to me appears erratic. Bats can pursue a flying insect at speeds ranging from a few miles an hour to a top speed of 100 miles per hour, often through dense forest—in the dark. And their wings are useful even when they are not flying. Bat wings are full of nerve receptors and are as sensitive as human hands.

Bats are also the only mammals that can fly long distances independently. About fifty million years ago, their ancestor, a small nocturnal mammal, began using its forelegs to glide through the air. Gliding gave this mammal access to the sky above and to the large food bowl of flying insects.

Bats were not the first animals to take flight. Birds and insects had been dominating the skies for millions of years before they got company. But while birds relied largely on light and vision to navigate, these new flying creatures had a different strategy. The bat's hearing enabled it to conquer a niche that most birds did not enter, namely the night. Bats and birds have, so to speak, shared night and day between them.

This conquest of the sky, and the dark, paved the way for new life-forms. More than a fifth of the roughly sixty-five hundred known mammal species are bats![2]

THE FACT THAT bats belong in the dark might explain why they have often been given a bad reputation.

My fantasies about bats resembling blood-sucking tennis balls had come from somewhere. In various myths and beliefs, especially ones from Europe and the United States, bats have been feared and disliked.[3] They have been viewed as conspirators with the Devil, and bringers of misfortune, melancholy, madness, and death. In some parts of Europe, it was thought that bats could turn people into thieves by sitting on their heads, which is a belief that endures to this day in some places.[4]

But their alleged connections to Satan are not the only thing bats have on their rap sheets. The Irish novelist Bram Stoker didn't do anything for their reputation when he published *Dracula* in 1897. In the novel, Count Dracula seduces young Lucy by visiting her in the form of a bat. But what has really put bats on the front pages in recent years is its role as a carrier of a coronavirus related to the one that triggered the COVID-19 pandemic.[5]

All of these things have likely contributed to my fear of bats. But I could have ignored my friend's older sister, and spared myself the anxiety, because there are no blood-sucking vampire bats in Europe. The small blood-sucking bats that live in South and Central America mainly drink the blood of livestock or birds. And most experts agree that, even if the coronavirus plaguing humankind does turn out to have come from a species of bat, it is our destruction of natural habitats, and our trading of wild animals, that is the real problem.[6]

We need bats: one small bat can eat three thousand mosquitoes per night. In many parts of the world, bats eat mosquitoes that could otherwise transmit malaria and other diseases.[7,8] There are also bats that pollinate plants and eat insects that could otherwise be harmful to crops. For example, it is partly thanks to bats that we can enjoy such delights as bananas, mangoes, and chocolate. It would be a bad idea to exterminate all the world's bats to avoid viruses.

A Landscape of Echoes

There's one! A movement in my peripheral vision caught my eye. Kjell and I had struggled back up the embankment and were standing on the floodlit path. A dark shape broke free of the shadows and swooped past one of the light cones radiating from the streetlamps before vanishing into the trees. I noticed how the bats' movements were somewhat haphazard, like a ricocheting ball in a pinball machine.

How bats manage to form an image of their surroundings, even in impenetrable darkness, has long been a mystery. Several scientists have tried wresting this secret from nature's grip, but the first breakthrough was made by the Italian priest Lazzaro Spallanzani in 1793.[9] Spallanzani must have had quite sadistic tendencies; he would experiment on bats by poking their eyes out with hot needles to see if they could still fly. He also filled their mouths and ears with glue to see if it affected their navigation. His contemporary, the Swiss physician Louis Jurine, filled bats' ears with starch. This substance clearly altered the poor animals' hearing so much that it disrupted their echolocation. The deaf bats became unable to navigate and moved around as if they were blind. But the principle behind echolocation remained undiscovered until just before World War II. In 1938, Harvard student Donald Griffin managed to demonstrate that bats create sounds at a frequency inaudible to humans. Because the quiet fall night down on the riverbank wasn't quiet for the bats.

In her essay "In Praise of Bats," author Diane Ackerman writes that bats "spend their whole lives yelling at the world and each other. They yell at their loved ones, they yell at their enemies, they yell at their dinner, they yell at the big bustling world." [10] It is a fitting description. A bat call can be as high as 120 decibels, which is loud enough to be painful for human ears. [11] Bats are therefore one of the world's loudest animals, so loud that their hearing organs are specially insulated to stop them from making themselves deaf. That humans are oblivious to these particular sound waves is perhaps just as well.

When a bat is flying, it calls eight to ten times a second. This call rate increases to about a hundred times a second when a bat identifies a bug it wants to eat and needs more detailed information. When these sound waves hit something—an insect, a raindrop, or a branch, for example—they bounce back and are received by the bat's ears. Based on the time it takes for the sound waves to return and whether they have been altered by their encounter with the surface they were reflected from, a bat's brain will construct an image of the landscape it is moving through. [12]

Humans use a similar principle to estimate how far a rock face is from us when we shout toward it, and then count the seconds until we hear the echo. Ultrasound images of a fetus in the mother's womb are also created by a form of echolocation. But while an image of a fetus has to be constructed by a computer program, the images a bat uses to navigate are created immediately in its head. Just as the comic-strip Batman's only superpowers are his intellect and strength, the bat's superpower is its brain and its ability to create images from an echo.

The bat's superpower, however, isn't entirely unique since it's one we have ourselves. In a hugely popular video on the internet, Daniel Kish, who is blind, cycles around with his eyes closed.[13] Your heart sits in your throat as you watch this so-called batman turning right and left, undaunted by the passing cars and buildings on all sides. How does he avoid crashing?

The answer is echolocation. Kish makes clicking noises that bounce back to him from the nearby surfaces. Like an echolocating bat, Kish perceives the world around him via sound. He has trained his brain to produce images of his surroundings by interpreting the echo, and is able to recognize the different surfaces and textures based on how they echoed the clicking sound. He "sees" with his ears.

BACK TO MY EVENING by the river. Kjell and I had come to one of the low bridges that span the water. Cars crossed the brightly lit highway just above our heads. It had been a while since I'd heard any clicking sounds in my headphones.

Kjell pointed at the iron construction overhead. It was cheerfully lit by purple and green lights, and looked like the twisted skeleton of a colossal whale.

"This could be a problem for quite a few bats," Kjell explained. Bats are safe while under the cover of darkness. Because while their flexible wings make them precision fliers, they're not fast enough to escape the predatory birds that are active during the day. Daytime, and light, means danger. When it's light, bats are easy prey. This fear of the light is a fear of being seen and eaten. In a way, it is a converse reflection of our fear of the dark.

The looming shadows that appear in our peripheral vision when our fear of the dark takes hold are the remnants of the lions and tigers and other very real monsters that plagued the humans of the past. It is our fear of the dark that makes us illuminate our streets, parks, and bridges. But it's the opposite for the bats flying up and down the river searching for food, or using the city's rivers as aerial thoroughfares. All these illuminated bridges and floodlit walkways become obstacles that are scary to fly past.

AT THIS POINT I need to make a small digression about attributing human characteristics to animals and other living things. During this year of exploring the secret life of the city, I have become attached to many of its creatures. I feel closer to the blackbirds, the ants, and the crows. I find that this empathy often happens when I gain insight into the types of challenges they face, challenges that are reminiscent of my own. The more I realize that they too become frightened and happy, the more interesting they become to me. Understanding animals as sentient beings has made it easier for me to feel an affinity for them.

Such anthropomorphism, when you project your own feelings onto something non-human, can be problematic.

We cannot be certain of what an animal is feeling. I cannot know if the crow felt some kind of schadenfreude as it dropped walnuts on the car belonging to my friend Geir Sonerud, the crow expert, or if a blackbird feels annoyed when it hears only fragments of the messages from its neighbor.

But there are some things we *can* know.[14] Some of the deep, instinctive, and affective neurological systems in animals and humans have the same biological origin.

I had by chance just been chatting with Knut-Petter Sætre Langlo, who is a psychologist and specialist in neuropsychology. I was curious to know whether human fear had anything in common with the fear that animals experience.

"The basic emotions are quite transferable between a lot of animals. Many of the same things are going on in the body." Langlo told me that many of the basic emotional systems developed early on in evolutionary history. An example would be parental love. Many mammals form deep bonds with their young. When scientists look at brain activity and hormonal systems in these situations, there is a great deal of overlap; the brains of different animal species react in a comparable way. When humans and bats want to protect their newborn offspring, they are governed by many of the same mechanisms. Fear is another example of a basic emotional system that is shared across animal groups. Most of the world's animal species get scared. It is a behavior that is governed by biological responses.

"Panic attacks are quite an illustrative example. They are a primitive bodily reaction in humans too." Anxiety may well be one such point where human and animal behavior intersects.

The crow researcher John Marzluff had shown, by scanning crow brains, that crows become scared in a similar way to humans. When the crows were exposed to a threatening face, it caused special centers in the amygdala and cerebral cortex to become active.[15] In other animals and humans, these same parts of the brain are associated with fear.

Our illuminated landscapes can have different effects on different species of bat. The most common urban species in Norway, namely the soprano pipistrelle and the northern bat, hunt in the light cones, despite the fact that it makes them more vulnerable to being caught by nocturnal birds. But many of the less-common bat species are too afraid of the light to join in. The lighting in urban areas could therefore lead to a shift in bat species composition, with the common species becoming even more common and the less-common species becoming rarer.

Of the thirteen species of bats registered in Norway, six are on the red list of endangered species. This is part of a larger trend: in recent decades, many of North America[16] and Europe's bat species have declined sharply as have populations in many other parts of the world. There are many reasons for this decline, but our floodlit towns and cities don't make life easier for many of these light-fearing creatures.[17]

As Kjell and I followed the path down toward the city, I tried to imagine how the terrain might look to a bat. Streetlamps were everywhere. In many places, large spaces between the trees and long sections of path were illuminated. I remembered the bat we'd seen earlier, how it had fluttered through the streetlamps, and I found myself thinking it was quite brave.

AFTER SAYING GOODBYE, Kjell turned and walked away while I continued walking downriver. Then I looked around to make sure I was alone and attempted something I hadn't done since I was little. I closed my eyes and took a few steps, as slowly as possible, while resisting the impulse to open my eyes.

This was something I did as a child, often on my way to school. I would try to beat my own record for the number of steps I dared to take before opening my eyes. Looking back, I remembered thinking it was a way of expanding my consciousness (I also realized I was lucky I didn't get run over). When I did it all these years later, it felt much the same: that when one sense was blocked, something happened to the others. Being on the footpath with my eyes closed was an entirely different experience to having my eyes open. I was suddenly aware of what sort of surface I was walking on. I could feel the mist clinging to my face like a veil of damp.

Then, with my eyes still closed, I tried to make a clicking sound by pressing my tongue against my palate. After a few attempts, I managed to make a noise that gave me an impression of the surrounding space. It could have been my imagination, but it felt like I could sense the trees lining the footpath. Either way, being deprived of sight wasn't easy, and I was relieved to open my eyes again.

The Smallest Among Us

All the beech forests that I know of look the same. They are built with the same architecture. A beech forest creates its own brown-green space. The layers of stiff, glossy leaves that make up the dense treetops absorb most of the sunlight before it reaches the ground.

But not all beech forests are as inhabited as the one we were in.

Bokemoa is an old beech forest in the middle of Stokke, a town in southeastern Norway. Although it is small and surrounded by housing estates, cars, and school buildings, the forest was fairly quiet when we visited. The beech trees stood with their smooth gray trunks and dense foliage, the forest floor was bare, and last year's beech leaves crunched underfoot. The piercing sunlight that had dazzled us as we walked through the streets on our way there was barely noticeable in the forest's cool and shady interior.

The "we" on this occasion was bat expert Magne Flåten and me. A few days after my bat walk with Kjell, I took the train out to Stokke where I was due to meet Magne at the railway station. Magne and I first met on a course where he gave a two-hour lecture on bats that ended with thunderous applause. Magne is retired, which is good because he has so many interests he would have no time to sleep otherwise. We stopped at his house for a coffee, where I was introduced to the art of origami bat making, before heading into the forest to look for the real thing.

We had come to a small clearing, and the sun was streaming through the open space in the treetops. Among the branches, just a few feet above my head, hung what could only be described as a little robot head, with an expressionless face and a narrow slit for a mouth. Several other identical heads hung from the surrounding trees, their robot eyes (two shiny screws) glinting in the sun.

"Here they are!" said Magne.

It then dawned on me that these strange constructions were in fact apartment buildings for bats.

The thing resembling a lower lip, Magne explained, was a landing platform just wide enough for an incoming bat. The horizontal slit, the robot's mouth, was the entrance to the box. Magne had hung up twenty-five of these bat boxes, creating a self-contained bat suburb in Bokemoa forest, among the schools and residential areas.

Magne picked up something brown and dry from the ground. "Bat poop," he said. "The proof that there are bats living in the box." He raised the ladder in his hands above his head and tapped the box with it.

"To check if there's a wasp nest in there," Magne explained in response to my puzzled look. Since no wasps flew out of the opening, he propped the ladder up against the tree and began climbing.

We had been searching for bats on the national Artskart map earlier that day. Magne had shown me that there are bats all over Norway, mostly in the south but also above the Arctic Circle in the most northerly parts of the country. And relatively dense clusters of dots—which indicate that someone has seen, heard, or caught a bat—existed in the towns and cities too. While Magne set to work unscrewing the robot eyes, which turned out to be holding the bat box's lid in place, we talked about how bats can fly right over our heads virtually unnoticed.

Bats spend their lives in two dimensions of the world that are largely inaccessible to humans—the sky and the dark. But also other facets of the bat's life seem exceptional from our human perspective.

To start with, bats are extremely small considering they are warm-blooded mammals. They are one of the *smallest* warm-blooded animals on the planet.

"A bat pup is as big as the phalanx at the tip of an adult little finger!" I checked my own little finger. Magne was right; it was pretty small.

Magne started to explain how he and his colleagues look after small bat pups that have become orphaned, by laying the pups on large pop bottles filled with warm water. Magne got so carried away while explaining how he fed them milk with a pipette that I was afraid he would fall off the ladder.

"They're so cute!" he laughed. "I get quite paternal when I think about them."

THEIR UNUSUAL SIZE made me wonder: Why *are* bats one of the smallest warm-blooded animals? Why aren't there mammals or birds as tiny as ants? Why can't we breed a Labrador retriever that can be carried round in a matchbox?

The reason is partly because small animals consume far more energy—per unit of body weight—than large animals do.[18] Small animals have a larger surface area in relation to their mass than large animals. They lose more heat relative to their weight than larger animals, and use more energy keeping themselves warm.

When a warm-blooded animal weighs less than around one-sixteenth of an ounce, its energy requirement per unit of body weight is astronomical. The bumblebee bat, which is native to Thailand and Myanmar, holds the record for being the world's smallest warm-blooded mammal.[19] It weighs about the same as a large bumblebee. I was unable to find any figures showing what this bat has to eat in order to meet its body's energy requirements, but the bee hummingbird, which is about the same size, has to eat half its own weight in nectar every day.[20] That's like a 130-pound human eating 65 pounds of food a day to keep up with their metabolism.

Norway's smallest bat species, the soprano pipistrelle, can be as light as three-and-a-half grams, not even a quarter of an ounce, which is about the same weight as three or four pieces of chewing gum. These tiny bats have an extremely large surface area compared to their body mass. The pups usually weigh less than a gram, about the weight of a single piece of chewing gum, and have to grow and gain weight, so they have an even greater energy requirement per gram of body weight than the adults. These bats operate at the very limit of how small a warm-blooded animal can be—in our cold climate at least. The amount of energy they require, per unit of body weight, is huge, which means they have to be voracious eaters, and their metabolism has to work as fast as possible.

Bat pups are born in the spring. "For bat pups, the summer months are a relentless fight to consume enough energy," Magne explained. It's a race against the clock because the pups have to gain enough weight to survive. To avoid starving to death during their first winter, the pups have to be eating machines during their first summer.

TEMPERATURE PLAYS A key role in the life of a bat because of its small size. And this is one reason why Magne and the other bat experts often get calls from people associated with Christian communities. Not because bats are especially religious, but because they have a penchant for church architecture.

Church spires with copper cladding, Magne said, absorb the sun's heat. The temperature on the inside of the metal roof can become hot enough to fry an egg. This microclimate makes them near-perfect summer colonies for some of the heat-loving bat species. And the very top of the spire, where the hot air accumulates, is where the pups often hang. The heat makes the enzymes in their digestive system work faster, enabling the little pups to metabolize more and gain weight quicker.

In favorable areas—such as those in southern Norway where there are deciduous forests—half of the churches can be home to brown long-eared bats, according to Magne.[21] "They seem especially keen on old churches that are spacious and have large entry holes, like the openings in the side of a bell tower."

Many bat species (especially brown long-eared bats) love living under flat roofs made of aluminum, tin, or copper. They will also crawl into chimneys that have been absorbing the heat of the sun all day, and the colony's young will lie there with their bellies pressed against the brickwork, suckling, digesting, and growing. This warm environment can allow a small bat pup to acquire just enough fat reserve to survive the winter.

A colony of bats is often a community of females and their young. The males have to leave once they become adults, but the females stay all summer, along with their pups and relatives. There can be many generations of bats roosting in one colony. In Norway, the largest recorded colony contained over a thousand bats, but a few dozen is more common.

In the forest of Bokemoa, where Magne and I were, the bats were living among joggers and dog walkers, with buildings and infrastructure on all sides. Even their homes were made by humans. I asked Magne where bats lived before the age of bat boxes, church spires, and sheet metal roofing.

He explained that when the bark starts to loosen from large old trees, a narrow pocket will often form between the bark and the wood. And when the tree is in the sun, these pockets can get very warm. They also provide relative safety from predators. The church spires and metal roofs are the human-made equivalents of these bark pockets.

A lack of safe and warm places to rear young is one of the biggest challenges for bat populations living near urban areas. In the cities there are normally long distances between old trees, which are felled to make room for roads and houses or because people are afraid of the wind blowing them down.

When bats find a good place to live, they usually hang onto it, staying there year after year, generation after generation. Because these places are currently so difficult to find, they become immensely valuable to the local bats. The pups' well-being is so dependent on a good place to live that if one is destroyed it is a catastrophe. That's why Magne and the other enthusiasts hang up bat boxes.

Bats usually give birth to just one pup per year. So, bats need to live a long time to produce much offspring. But what also makes bats so exceptional is their lifespan.

Most small mammals have relatively short lives. But bats can live for a long time. So long, in fact, that scientists aren't quite sure how old they can get. Magne talked about how bat researchers in Siberia had found a small male bat that was at least—and possibly well over—forty years old. It's not always easy to determine the age of a bat, he explained. "When bats are adults, it's almost impossible to find out how old they are. But this particular bat had been tagged by scientists. And it was already an adult when it was tagged, although no one knew how old it was at the time. Then, forty-one years later, it was found again."

A bat's teeth can provide a clue as to how old it is. Mastication over the lifespan of a bat wears down the tooth surfaces, and the degree of tooth wear can be used to estimate how long it has lived. The bat found in Siberia, which was over forty years old, had no sign of wear on its teeth. Another bat, also found by Magne and his colleagues, had teeth that were completely worn down. We can only wonder how old that one might have been.

On average, a bat lives about three times longer than other mammals of the same size and metabolic rate. Bats are nervous creatures, Magne explained, and this might be one reason behind their longevity. "If you only weigh a sixth of an ounce, and are surrounded by enemies, you won't live to age forty without being a little cautious! Being nervous has been a good survival method for almost fifty million years."

This innate nervousness is the reason why people who have bat boxes often have to be patient to see bats inhabiting them. Magne said that it took a long time before the bats moved into the robot heads all around us. A bat mother won't move into a box with her pups until she has watched it for some time and is sure that there are no predators or other dangers lurking around. "Perhaps several years," laughed Magne.

IT MADE SENSE. Bats are able to share urban spaces with us, without being noticed, because our paths rarely cross. They avoid the light, and are constantly airborne. They sleep during the day and hunt in the dark. They are extremely cautious, and very small. This combination of factors may account for why we allow bats to remain a mystery. But our worlds have to overlap sometimes. Even bats need to rest. And that's why I was standing in front of the robot head with my heart in my throat.

Magne had stopped what he was doing while we talked and was finally loosening the screws on the bat box. Was I about to come face to face with a bat?

The box was empty.

Each bat box contains thin partitions with horizontal grooves carved into them, like a 1990s CD shelf in miniature, from which the bats will hang by their feet when resting. Just one of these boxes, which is no bigger than a three-pint milk carton, can house an entire colony: dozens of animals—mothers and pups—all crammed together. On the ground below, small piles of droppings indicated that bats had used the boxes during the summer. But the occupants had gone.

I was feeling a bit disappointed as Magne drove me back to the railway station. It was as though I hadn't quite made contact with the bats. As though they existed in a parallel universe—a world similar to ours, but it's mirror image. Our two worlds share very few points of contact, and I had failed to find any of them.

Those Who Sleep

The sound of footsteps echoed down the narrow mineshaft. Beams of light from the headlamps bounced and danced in the darkness. It was early January, the biting temperature outside had dropped below freezing, and inside the mine the temperature was only just above zero. About a dozen of us were making our way, one by one, across the narrow wooden planks that had been laid over the puddles, into the vast network of tunnels that penetrates the mountain. We had been told to be quiet. Because the tunnel's residents were sleeping, and no one likes to be woken up by strangers.

I was there because the Norwegian Zoological Association's bat group wanted to count the hibernating individuals in an old cobalt mine a couple of hours' drive from Oslo. They had been doing this count for many years, to help find out what bats do during the winter months, because it is something experts know very little about. Compared with the number of bats flying around on a summer's night, very few have been found in the winter.

It is assumed that while some species migrate south, most of the Norwegian species probably hibernate in

Norway, in places that are humid (these little animals get easily dehydrated) and where the temperature stays just above freezing. They will fly into caves or mines or in between the large rocks on a talus slope. Then they hang there, in deep hibernation, breathing once an hour as their hearts beat once a minute.

Magne had said that bats are extremely precise when it comes to temperature, just as they are with most aspects of their lives. "It can only be a few degrees. If it gets too warm, their metabolism will speed up. They'll then start burning their valuable fat reserve and starve to death before spring. If it gets too cold, they risk waking up with frostbitten ears, or not waking up at all."

"Bats are extremely good planners," Magne said. "They will have at least a plan A, a plan B, and a plan C for where they are going to spend the winter." According to Magne, bats can spend weeks during the fall looking for the best places to spend the winter. They will know of several locations, some deep and some shallow, that they can move between while they are hibernating, depending on how warm or cold a winter it is.

"In the fall the bats virtually stop eating. They just fly back and forth, in droves, between potential hibernation sites. All bats engage in this swarming. They have to check, over and over again, that these places are nice and safe. There is such a lot at stake!" he said.

"How they manage to do this is a mystery. But they follow their grandmothers and great-aunts, learning where the best hibernation sites are from their elders. And it pays off. The winter mortality rate for those who find a safe place to hibernate is low."[22]

According to Magne, what *has* caused problems for hibernating bats is something that humans invented: pesticides. These chemicals can have a direct effect on bats, via the insects they prey on. "The insects may have ingested toxins from the plant nectar they have eaten."

Fat-soluble toxins are stored in body fat. During the summer and fall, bats continually gain weight and so their toxin levels don't increase. But when they are hibernating and using up fat, the toxins accumulate in the only fat-rich organ they have left: the brain.

In a study from Indiana, nearly all of the 332 bats tested had traces of organochlorine pesticides in their bodies.[23] Many of the pesticides the researchers found hadn't been used for decades. The consequences, for mammals, of ingesting these types of pesticides can vary. They can cause problems with reproduction, with eating, and in the worst case, they can cause death.[24] Bats can die from poisoning when they are asleep, while hanging upside down in their caves.

I hoped this wasn't the scene waiting for us inside this well-known bat hibernation site in eastern Norway. Still, a degree of tension hung in the air as we entered. The group leader, Kristoffer Bøhn, gave us some hand-drawn maps and instructed us to count and determine the species of bats we found. We had to look thoroughly in all the cracks, while also ensuring the batteries in our lights didn't go flat, and above all else, that we didn't get lost.

The impenetrable dark beyond the beams from our headlamps was ice cold. Bats are safe from most dangers in these conditions. No animal that depends solely on its ability to see would find its way here, deep inside the mountain, unless it was wearing a headlamp. But by using echolocation, bats can find small holes or cracks just big enough to squeeze into, where the darkness protects them.

AN HOUR WENT BY without us finding a thing. I was starting to resign myself to the fact that bats would remain no more than movements in my peripheral vision.

But wait!

A speck of dense, short fur interrupting the rugged surface of rock made me jump. Just inside a borehole, I saw a pair of ears in the light from my headlamp. I aimed the beam to one side and held my breath to stop the vapor clouding the light field. And there it was. A tiny thing, hanging motionless, its head pointed downward with just a snout and a pair of large translucent ears poking out of its soft gray fur. It was hard to understand how a heart could be beating in there.

Like a two-and-a-half-inch stalactite, the bat hung peacefully with its eyes closed. One of the city's most mysterious creatures. Finally, here was the bat I had been looking for, up close. It looked more like a tiny puppy than a blood-sucking tick. I began to understand Magne's statement about bat puppies making him feeling paternal. But at the same time, it felt like an intrusion to be so close to something that had gone to such great lengths to hide, to be spying on something that was certain it was alone.

This was the closest I would get to a bat. But I knew that by the summer I would once again see bats fluttering around the streetlamps and go looking for their droppings under the bat boxes. I would borrow a time-expansion detector and listen once again to what the city sounds like to a bat's sensory apparatus. I took a quick photo of the bat, pointed the light from my headlamp away, and left the bat hanging there in the dark.[25] Just knowing that it was there was good enough for me.

The Written Language of the City

A YELLOW-ORANGE SPOT caught my eye. The spot was about two inches across, with orange streaks radiating from the center like strokes of thick oil paint. It was November, and almost a year had passed since the orange structures at the Antarctic penguin colony had seemed beautiful and exotic. Nevertheless, until this moment I had failed to notice the spot on the cherry tree next to the bike rack in the yard.

I thought about it for a few seconds, then turned, walked up the stairs, and went into my apartment. In a bag of field-trip equipment that had been packed away in the loft for the winter, I found my magnifying loupe. The loupe's light and its twenty times magnification brought almost anything into extremely sharp detail; the skin on your thumb, for example, looks scaly like a reptile under the lens. I ran back down the stairs and out to the cherry tree, then reached up to the yellow spot with my loupe while leaning against the trunk.

Under the lens, the lichen was as strange and beautiful as it was the last time I saw it up close. Its texture was like solidified lava, with tiny orange bowls stretching up from its gnarled and craggy center. Some of the bowls were flat and plate-shaped, some were more like soup dishes, and others had stems like crudely fashioned wine glasses. The surface inside the bowls had a smoother, almost glazed texture. Viewed so closely, the lichen looked almost like a dinner service made by a wildly eccentric ceramic artist.

I was quite surprised by how I'd missed this spot until now, considering I'd actually spent part of my time as a biologist specifically looking for lichens. I had searched in old pastures and old-growth forests, places where you can find significant rarities. In the forest, lichens were often everywhere. Large bushy clusters of gray-brown strands that hung like beards from the branches and clung to your hair when you tried to get past. Exposed rocks covered in bushy lichens that crunched underfoot when it was dry and felt soft like sponges when it was wet. And deep within cracks in the bark of the really old trees, you would find pin lichens: tiny, crooked stems with a little head on top.

What had fascinated me most about lichens was their ability to tell us about the environment they are living in. Many lichen species have strict requirements when it comes to their habitat. They will grow only in places that have a certain level of humidity, that get the right amount of sunlight, or that have remained fairly unchanged for a long time. Lichen composition can indicate, for example, how deep the snow level was last year or how old the forest in an area is. It can tell us what an area's climate is normally like, how long a pasture has been used for grazing, and how long a tree has been left undisturbed. These are fairly approximate estimates, but still. If I had to give lichens a human characteristic, I might use the word "sensitive."

But I had learned how to read lichens in a *non*-urban environment. So while I stood beneath the cherry tree, I started to wonder if urban lichens have something they want to get off their chests too. Why was this lichen growing right there?

I couldn't find a good answer. But I did start to carry the magnifying loupe with me during my daily errands. I poked my nose into broken bark and peered up at the treetops in the streets and parks. The beautiful forest lichens were nowhere to be seen. There were no white bushes covering the roofs and walls, no clusters of witch's-hair lichen hanging from the branches, and no tiny heads of pin lichen in the cracks of the large trees standing in the parks.

What I did find were flakes of white, brown-green, or yellow-orange lichen growing on walls, tree trunks, and branches. It was as if the remarkable three-dimensional textures of the forest lichens had been flattened on their way into town.

Gossiping Lichens

Einar stood with his nose pressed against the trunk of an old Norway maple. He was doing a routine search of the cracks in the bark with his loupe. We were in the botanical garden, and the maple we were standing beside had almost no lichens on it.

"Asthmatics might feel a certain kinship with lichens."

A couple pushing a stroller stared at Einar for a while as they trundled past. And he must have noticed them somehow because he smiled while leaning against the tree, interrupting his own train of thought.

"The trick is to act like nothing's going on," he said. "People always stare. You can't let it bother you." Einar tore himself away from his intimate encounter with the maple tree and brushed his jacket for bark flakes.

Einar Timdal, associate professor at the Natural History Museum in Oslo, had just returned from a field trip to the Amazon. He has endured worse things for the sake of lichens than having people think he is weird. We had met outside an old brick building in the botanical garden. Above the building's ivy-clad entrance a sign read "BOTANICAL MUSEUM" in capital letters.

Einar often goes round wearing '70s band T-shirts under his Gore-Tex jacket. He's the sort of guy you'd suspect might have a hefty sound system at home. He can often be seen surrounded by flocks of students, with a loupe around his neck, a knife in his belt, and a small hammer in his hand to tap loose the lichens he finds growing on the stone. I had been one of those very students who'd asked the same things repeatedly. But Einar was always patient and could answer just about any lichen-related question.

He had now taken a step back and was in a more vertical position. His magnifying loupe dangled from a red cord that hung around his neck.

"It's about air. The first studies which showed that lichens respond to air pollution came from the British Isles as early as the 1860s," he said.

I tried to imagine London in 1860 and saw streets full of Jack the Ripper–like figures: men in top hats and long coats, women wearing crinolines, horse-drawn carriages. A thick yellow-gray pall of soot and sulfur dioxide shrouded the cobbled streets. In his dictionary of London from 1879, Charles Dickens wrote: "During the continuance of a real London fog—which may be black, or grey, or more probably orange-colored—the happiest man is he who can stay at home..."[1] The greasy smog that made Charles Dickens' eyes sting was a combination of smoke from the hundreds of thousands of coal and wood stoves, and the fog from the Thames.

IT WASN'T JUST DICKENS who disliked the environment in London at the time. Lichens don't have roots; they absorb water and nutrients straight from the air. So, if the air is full of sulfur and other toxic substances, those toxins end up being absorbed into the lichens. It's the equivalent of our own bodies being covered in lung tissue, instead of skin, which would allow any soluble substances we came in contact with to go straight into our bloodstream.

Lichens are slow growing and can live for a very long time. An Arctic map lichen, native to Lapland, Sweden, is the oldest known specimen of lichen and is estimated to be nine thousand years old.[2] Not all lichens live anywhere near that long, but they often live long enough for harmful substances to accumulate in their vegetative tissue.

The first lichens to vanish from London's increasingly smoke-filled streets were the bushy lichens. These lichens, for example the witch's-hair lichen that hangs from branches in the forest, have a large surface area in relation to their body mass. The large surface area makes bush-shaped lichens better at absorbing gases than flat lichens. This is a generally positive trait in environments where nutrition is in short supply, but it quickly becomes deadly in polluted cities. The large bushy lichens are the first to disappear from the cities when air pollution increases.

The lichens we see stuck to a wall or tree have a smaller surface area in relation to their mass, and therefore they tend to last longer. But in the middle of the increasingly smoky cities, the air pollution levels eventually became so high that often no lichens grew there at all—neither flat nor bushy ones. London, for example, became a lichen desert in the mid-1800s.

But one lichen lasted longer than the others: *Lecanora conizaeoides* looks like wrinkly, green-gray skin covered in pale-orange warts. "It's called city rim lichen in Norway because it's only found in our cities," said Einar.

This lichen probably originates from the hot springs of Iceland and can cope with the sulfurous fog created by Iceland's geothermal heat better than most other organisms.[3] Atmospheric sulfur dioxide, whether it comes from hot springs or from polluting industry, has an acidifying effect on many of the surfaces where lichens would like to grow, such as a tree's bark. It also creates more acidic rain. When England's cities with their smog and sulfurous mist expanded, the sulfur-and-acid-loving lichen suddenly acquired a whole new habitat. Due to a coincidence of nature, it had already adapted to living in sulfurous smoke and acid environments, in roughly the same way that rock pigeons had been ready to live among the high-rise buildings and asphalt that emerged in cities.[4]

Although originally unique to Iceland, the city rim lichen was first recorded in England in the 1860s. And as air pollution increased, this lichen spread rapidly to cities all over Europe.[5]

For most organisms, sulfur vapor is not particularly beneficial. Smog caused severe respiratory problems among humans, and the urban mortality rate increased to absurdly high levels when the pollution was at its worst. Even the city rim lichen died when the concentration of sulfurous gases in the air became too high. "That's why we often find a ring of city rim lichens around the most polluted area, usually right in the city center where it is totally dead," said Einar.

Perhaps the city rim lichen can be interpreted as a signal to potential home buyers. Can you handle more sulfur than the city rim lichen? Buy downtown!

THE FACT THAT different species of lichens tolerate pollution in different but predictable ways proved handy: in the 1960s and '70s, British researchers developed a scale showing which lichen species had disappeared and which sulfurous gases were to blame.[6] This meant that ordinary people could determine the degree of air pollution in their neighborhood. All you needed was a loupe, a lichen guide, and some patience (learning the differences between the various species can take a while). You would then find the nearest tree, check which lichens were on it, and look up the corresponding value of sulfurous gas: "No tube lichen, you say? Then the pollution level is probably over 70 micrograms of sulfur dioxide per square meter. Not even city rim lichen? Oh, then we are up to 150 micrograms per square meter. Perhaps it's time to move house, dear?"

In 1970s Britain, fifteen thousand schoolchildren took part in a survey of the lichens in their neighborhoods. This voluntary campaign provided British researchers with the basis for a map that showed the levels of air pollution across the British Isles.[7] The map closely matched the measurements that more advanced techniques brought to the table. So, long before citizen science became a buzzword, lichen surveys were a way for ordinary people, with no access to expensive equipment, to gain some control and knowledge about their local environment. Low-tech environmental monitoring, basically.

The lichens in British cities disappeared one by one as the air became increasingly polluted. But what happened in Norway?

"Air pollution has been bad in certain areas of Norway," Einar said. We had left the large maple tree and were walking east through the botanical garden toward the busy ring road and the noise of traffic. Einar told me that in the early 1900s Oslo was affected by the smoke from burning wood and coal, and sulfur emissions from factories smothered many of the cities there too. Just like in Britain, the lichens disappeared one by one, first on the outskirts of the city and gradually toward the center.[8] Although air quality has improved in a lot of places, lichens are still disappearing, one by one, toward the most polluted areas of many Norwegian cities.

LECANORA CONIZAEOIDES, the city rim lichen, is no longer the hardiest lichen in town. Our cities have now been taken over by another lichen, the nitrogen-loving *Xanthoria parietina*.

We stood at the fence on the east side of the botanical garden while the buses and trucks thundered by on the ring road twenty yards away. I had turned on my lichen vision and was eagerly scanning the area.

A row of young trees with yellowish trunks lined the fence. And as we got closer, we noticed how their bark was covered in round patches of yellow-orange lichen. This was a huge contrast to the almost bare maple bark we had looked at nearer the middle of the park. Einar took me to a tree marked by a plaque that said "Swedish whitebeam." A few brown, lobed leaves swayed dolefully in the breeze. Clusters of berries similar to rowanberries hung from its narrow branches. The trunk was covered in yellow-orange, green-brown, and gray patches of lichens, small communities that expanded in a circular form until they reached the very edge of the next community. Einar placed his loupe against the tree, and the species names just rolled off his tongue. These were Latin words that you instinctively knew were meant to be in italics: *Phaeophyscia orbicularis, Melanohalea exasperatula*.

Halfway up the trunk I recognized the yellow-orange color and the lava-like texture of the lichen in my backyard. "This is *Xanthoria parietina*," Einar said, pointing at the patch of lichen.

I HAVE TO PAUSE for a moment. The Latin name for the *Xanthoria* genus to which these lichens belong is derived from the Greek word *xanthos*, which means "yellow." However, these lichens can also have a more orange or red hue.[9] It was this genus of lichen I had found among the Antarctic penguins. In Norway we have around ten species within the *Xanthoria* genus. Common orange lichen, *Xanthoria parietina*, is a rosette that attaches itself to trees, roofs, fences, and other surfaces in Norway's lowland areas. It grows slowly across the substrate; the growth rate varies but it is usually between one-sixteenth to one-eighth of an inch a year.[10] So the circle on the cherry tree in the backyard was certainly a few years old.

I moved a little closer and picked off a bit of the yellow-orange lichen. Its underside was grayish white and covered with coarse hair. Its top side was covered in tiny bowls with orange bases, especially toward the middle of the lichen. "The English call them fairy cups," Einar said.

What could the common orange lichen tell me about my backyard? And about the mountain ash growing beside the ring road?

The emissions of sulfurous gases that made London an ideal habitat for the sulfur-loving lichens of Dickens' time are fortunately no longer such a big problem in most cities, Einar explained. Today, the leading pollutant in many cities is nitrogen rather than sulfurous gas. And that's where the orange lichens come into the picture. These lichens can endure frigid temperatures and harsh climates, but what they must have is a good supply of nitrogen.

"The reason you often find orange lichens around nesting cliffs," Einar said, "is because they like the bird droppings that are rich in nitrogen. Small particles of these bird droppings are also carried by the wind to places all over nesting cliffs. The landscape can have an almost orange hue in these places."

I thought of my time working as a guide for cruise-ship tourists in the polar regions. The ornithologists on those trips could spot a nesting cliff before they actually saw the birds. Now I understood that they had spotted the cliffs from a distance because of their color.

Over the last two hundred years, access to nitrogen has increased dramatically in many places. Leaked fossil fuels and fertilizers have added huge amounts of nitrogen to the global cycle. Vehicle emissions, especially those from diesel cars, are a major source of nitrogen oxides. And this increase is very noticeable in cities.

"What's happening now is that many of the lichen species that actually belong in nitrogen-rich environments, along the coast for example, are now taking over the cities," says Einar. The written language of lichens is spelling N I T R O G E N all over the city.

While Einar and I stood at the entrance to the botanical garden, I asked him if he agreed that lichens are a form of written language. And he did. "It's fun keeping track of how lichen communities change," he said. "I can't help looking at old trees, even when I'm on a night out on the town with my wife." Einar's interest in lichens had taken him around the world. But not that many people feel so passionately about them. Countless blogs and Facebook groups are dedicated to plants and flowers. Birds are popular too; even mushrooms have a fan base of biologists and amateur mycologists. Lichens, on the other hand, have fewer fans. They are easy to overlook. To my knowledge, no one has ever written a poem about lichens, no one's betrothed has been compared to a ripe *Lecanora* or a voluptuous fruiting body.

But two days later, while driving along the ring road, I suddenly noticed that the tree trunks beside the road were orange while the trunks about thirty feet from the road were smooth and gray. Not exactly poetry, but still. It felt like I'd learned to read a few words of lichen script.

P.S. Lichen in Space

In 2010, a small piece of a species of orange lichen, namely the elegant sunburst lichen, was placed outside the International Space Station as it orbited Earth. For eighteen months that lichen was exposed to cosmic radiation, a vacuum that causes dehydration, and extreme temperatures. Not only did the lichen survive on the outside of a spacecraft, it also managed to perform photosynthesis and actually *adapt* to the new conditions in space.[11]

Fungi, bacteria, and other organisms were also similarly tested, but none of them survived anywhere near as well as the lichen.

Why is lichen so resilient? Lichen is a cooperative organism, a symbiosis between a fungus and an alga. If you cut an orange lichen into thin strips and put one of these strips under a microscope, you will see that in cross-section the lichen looks like some kind of Napoleon cake: the puff pastry sheets are formed by a solid net-work of fungal hyphae, while a green and porous filling is formed of algae cells and loosely woven fungal hyphae. The algae perform photosynthesis while the fungi provide the algae with good living conditions. Fungi have a wide range of chemical and physical tricks up their sleeve that enable them to survive and protect the algae they are car-rying. Meanwhile the algae can continue to produce sugar, safely enveloped by the fungi's protective atmosphere.

It has been a winning formula: lichens cover about 8 percent of the planet, an area equal to all the world's rain forests combined.[12] And lichens can live in space.

Researchers are now looking at how fungal and algal cells manage to avoid radiation burns or fatal freeze-drying in space's extremely low temperatures. They hope to understand how life can arise in extreme conditions, and how cells manage to protect themselves from radiation and vacuums. Some researchers believe the cooperation between fungi and algae is precisely why lichens thrive so well, in both space and the extreme conditions of Antarctica.

THE FACT THAT a lichen can survive long periods in space has bolstered supporters of the (rather speculative) pan-spermia theory.[13] This theory states that life—in the form of extremely resilient lichens, for example—may have spread between planets, on asteroids perhaps, and can therefore be found in several parts of the universe. Could lichens have been the first life to arrive on planet Earth? Perhaps they will manage to spread from Earth to other planets?

This idea is pure speculation, of course. But the experiment has shown that it is theoretically possible for relatively complicated organisms such as lichens to survive space travel after all. Lichens' combination of both fungal and algal cells would give them two base materials with which to kickstart their life on a new planet. It would also provide enough material for *Science Illustrated* and other popular science magazines to write articles about for decades to come. Whatever the case may be, I certainly won't make the mistake of overlooking lichens again.

The Near
and the Dear

THE SPARROW COCKED ITS HEAD and stared at me. Then it hopped even closer to the cinnamon bun on my plate, until it was so close that I could see the tiny feathers around its beak fluttering in its breath.

In the square where I sat at a café table, the Christmas market was underway. A horseshoe of stalls piled high with knitted socks and jars of honey arched beneath the trees, while vendors competed over who could stock the most locally produced and heartwarming gifts. People looking for a Christmassy atmosphere queued here and there with steaming paper cups in their hands. It was above freezing, the snow had melted, and the air was loaded with the smell of fresh waffles and wet asphalt. I sat at a café table wrapped in a woolen blanket. The few rays of sunlight that caught my face were not enough to warm my skin. The December sun had spent half the morning rising, and the opalescent patterns in the sky indicated that it was already on its way back down.

I moved my plate with the cinnamon bun to the seat beside me. Too late. A house sparrow glanced down at the plate, then at me. Had we been alone I would have gently swatted away the bird with my newspaper, but I didn't want to alarm the families sitting nearby. Besides, I was in the minority. Hundreds of other sparrows flitted about in the square, hopping around the bicycle wheels and children's feet. The flock gave off a constant and almost plastic-sounding chirp. *CHEEP CHEEP CHEEP.* Even while they were eating. I wondered if they ever shut up.

Watching the sparrows was like watching a shoal of fish. They all moved as one, which made it hard to keep your eyes fixed on an individual—except for the bird standing before me. That bird was the vanguard of the flock.

Other than the gift of impudence, the sparrow in front of me had very few distinguishing features. It was a female, with a fairly large head compared to its body, which was a gray-brown color. And when it stretched its neck to get a better view of what was on my plate, it went from almost ball-shaped to being surprisingly slender.

The bird retracted its neck and returned to being round. Its plumage was made up of varying shades of brown. Of course, through a sparrow's eyes the female may well look as patterned as a carnival queen. A bird's retina contains four types of visual cells whereas a human's has only three. Many bird species can therefore see the short-wave, ultraviolet spectrum of light. As we watched each other, me and the bird, I wondered what her ultraviolet world looked like.

The sparrow and I sat looking at each other. I had seen its kind thousands of times before. But I had never before noticed the light stripe that went from the rear corner of its eye toward the neck, like the arms on a pair of glasses. Its legs were pale orange, thin as pencil strokes, with tiny raptor-like folds of skin that are a throwback to its dinosaur ancestry. I was close enough to see the details of the dirt-brown down on its head and breast, which gave the sparrow its disheveled appearance. Its wing feathers, on the other hand, had been carefully preened.

The bird hopped down onto the chair, then blinked. A whitish, almost transparent film appeared from the corner of its eye, slid round its eyeball, and just as quickly disappeared. I was startled, unaccustomed as I was to seeing a bird's inner eyelid up close.

It's difficult to argue with a mini-dinosaur. So, I stood up, grabbed my coffee, and left the half-eaten pastry on the chair.

Two seconds later I heard a roar of fluttering wings and turned to see sparrows flying in from all directions. The pastry was immediately knocked to the ground, where it vanished under a mass of pecking beaks, like a pack of hyenas in a feeding frenzy on the savanna. It was great entertainment, even without the snappy editing, dramatic music, and narrator's voice of a TV nature program. I stood there for a while, marveling at the chaos of wings and beaks, then went on my way.

The Companion

I walked down toward the river while thinking about how something had changed over the past year. Well, the sparrows, ants, and lichens were the same—but something had happened to *me*. I was less reluctant to stop for five minutes to watch a sparrow or some ants or the cracks in a tree trunk. And I was surprised to find that it didn't make me five minutes late to every appointment.

As the low rumble of rushing water slowly replaced the hum of the market, I thought about the sparrow on the café table. "The sparrow is an extremely sedentary bird," my ornithologist friend Bjørn Olav Tveit had said during the bird-watching trip we'd been on in central Oslo a few months earlier. We had passed a flock of sparrows chirping loudly in a bush.

The house sparrow originally came from Eurasia, and has since spread to Africa, Australia, and America, often more or less intentionally with human assistance. Bjørn Olav explained that although sparrows have succeeded in traveling all over the world, it's not because they are particularly adventurous. On the contrary. They'll often get trapped in ships' holds, for example, and cross the world's oceans as stowaways. They invade continents because they're too afraid to fly across the open sea once the boat has set sail. They are the anti-heroes of conquest.

I'd asked why the house sparrow is particularly common in our cities. "It's got a strong stomach that can handle all the shit we humans put in us," Bjørn Olav had laughed. "The sparrow's evolutionary strategy is to live on fast food!"

Maybe he's right. Research has shown that the urban house sparrow's diet contains more fat than the diet of a sparrow that lives in a more rural area, and that urban house sparrows have higher cholesterol levels than those living in the countryside. Living on human leftovers comes at a price. And the sparrow's ability to change its eating habits has been one of the things that has enabled it to live with us.

ON THE STEPPES of central Asia, a shy little bird called *Passer domesticus* ssp. *bactrianus,* the bactrianus sparrow, can be found hopping about. This sparrow subspecies lives a secluded existence. It is very similar to the house sparrow we know but is afraid of humans, just as most other small wild birds are. Instead of cinnamon buns and other refined wheat products, it lives on seeds and insects.

A group of researchers, some of them from the University of Oslo, traveled to the Chokpak Pass, an area roughly four thousand feet above sea level in southern Kazakhstan. It is a beautiful green landscape surrounded by high mountains. This is the bactrianus sparrow's habitat. The researchers went to a virtually dried-up riverbed and set up large funnel-shaped nets. When a bird flew into one of the nets, the researchers would pick it out, and, while holding it carefully, take a blood sample before releasing it.[1] By analyzing the blood samples, the researchers were able to compare the genes of the bactrianus sparrow with the genes of the urban house sparrow. The scientists looked at the genes that affect the breakdown of starch and those that affect skull shape. Their findings became a key part of the story about why the house sparrow is now such a regular sight around our café tables.

The researchers' data suggest that the present-day house sparrow and bactrianus sparrow are both descended from an ancient bird that would have very much resembled *today's* bactrianus sparrow. As in, somewhat timid and predisposed to eating seeds and insects. At some point or other, the two birds went their separate ways. This divergence probably occurred about eleven thousand years ago, in the Middle East, just before agriculture emerged there.[2] Some of these sparrows must have thrived around human settlements. Perhaps they gorged themselves on spilled or unattended grains. As agriculture spread to more and more areas during the Neolithic revolution, it brought the increasingly tame sparrow with it.

Wild birds and animals don't suddenly switch to eating starchy foods like grain and processed farm products overnight; their digestive system has to adapt. And scientists studying our feathered urban cohabitants found signs of this change in a gene associated with their ability to split and digest starchy grains.[3]

Similar changes in the genes associated with an ability to break down starch have been found in dogs, humans, and other animals that have gone from surviving on a wild to a more cultivated diet during our transition to an agricultural society.

The house sparrow is probably the most widely distributed wild bird species in the world. Following in human footsteps was clearly a good strategy.

BUT HOW DID the house sparrow that sat on my café table become so fearless?

Researchers can measure a bird's tolerance for humans by its flight distance; that is, by how close a bird will allow a human to approach before it flies away. A timid bactrianus sparrow will often fly away long before you get near it, just as other wild sparrows do. By comparison, the flight distance of the urban house sparrow, according to my calculations, is about four inches.

It's possible that the house sparrow's cheekiness is directly related to the fact that it cohabits with humans. The urban sparrow conquered the world hot on the heels of humans, while its timid cousin was left behind on the steppes. The same thing happened with many other birds that followed humans into the cities. In a study from 2016, researchers looked at bird species that have populations both inside and outside urban areas.[4] They found that the city version of the birds had a generally lower flight distance than the birds living outside the city. We get closer to these birds in urban areas than we do in non-urban areas. And not only that: the species that had lived for the most generations in urban areas were the least frightened, which suggests that the changes are genetic.[5]

Anyway, back to the house sparrow. Its Latin name, *Passer domesticus,* roughly means "small bird that belongs to the house." Researchers at the University of Oslo refer to this bird as one of humankind's companion species.[6] And as I walked by the river—with one less cinnamon bun than I started with—I thought about how rare it is for a biological expression to be so appropriate. The word "companion" is derived from Latin and literally means "one who breaks bread with another."

And this arrangement doesn't just benefit sparrows, I thought as I warmed my hands on my paper coffee cup.

Humans and sparrows have coexisted through some of recent history's greatest upheavals, for both species. But the fact that sparrows play an important role for humans is perhaps something we appreciate best when they're gone.

IN LATE 1950S, Chinese leader Mao Zedong introduced the Four Pests campaign,[7,8] encouraging people across the country to kill sparrows. Admittedly, this wasn't our own house sparrow *Passer domesticus* but a close relative, namely the Eurasian tree sparrow, *Passer montanus*. In Europe, we also regularly share our café tables with this tree sparrow, which has a similar way of life and diet to the house sparrow. The goal of the Chinese sparrow massacre was to stop them from eating agricultural crops.[9] And so the people of China destroyed sparrow nests and eggs, and beat drums to prevent the birds from landing. The sparrow populations collapsed.

The main problem was that sparrows also eat insects. And the extermination of sparrows led to a significant rise in insect numbers, especially grasshoppers, which were soon eating more grain than the sparrows had done. Instead of the harvests being increased, they were sharply reduced.[10] The campaign led to a prolonged ecological imbalance and contributed to the Great Chinese Famine, which claimed tens of millions of lives.[11]

Nature Without a Manual

The path I was on ran alongside the river, up toward the forest. The surrounding valley became narrower; its steep slopes blocked out the city, and the trees created a distinct space of their own. I could vaguely see a couple of apartment towers through the leafless branches. An illuminated footbridge cut across the path, and when I passed under it I heard a constant drip of water echoing around the bridge supports. I then recognized where I was. It was here that we had seen bats swooping through the streetlamps in the fall.

I paused and sipped my coffee while wondering how the bat in the cobalt mine was getting on. Perhaps the mild weather had penetrated the mountain, raising the temperature in its tiny furry body a little, maybe even enough to wake it from its deep sleep and make it wonder if spring had already come. I imagined it stretching its leathery wings. Then yawning perhaps, before wrapping its wings round its body and falling back to sleep. Maybe I'll see it flying between the office towers next fall, I thought. This time, however, I'll know who's careening in and out of the light from the streetlamps. I felt excited about the prospect of seeing the little furry stalactite again.

I turned and started walking downriver, toward home. The return journey offered a view of the Oslo skyline. The sky was pink, and the short December day would soon be over. In the shade of the trees, beyond the light from the streetlamps, the contours began to dissolve. The black river snaked beneath the willow branches. Circular patterns on the surface, which looked like rapidly blooming ice flowers, indicated where the river currents forced the water up. The mallards had not yet fallen asleep; they just bobbed about along the shore like pudgy wine corks.

On a small flat area near a bend in the river, I stopped and listened. A familiar chirruping noise was coming from a large cluster of thorny bushes on the riverbank, a non-native bush, an alien species that had spread without permission and was therefore hated by many. However, these bushes were not a significant problem for sparrows. Amid the thorny branches were hundreds of birds, and it seemed like every single one of them was constantly chirping. Were they talking to each other, I wondered? I took a step closer and the chirp changed slightly: a warning being sent from bird to bird perhaps, about a possible two-legged menace outside the bush.

The house sparrow must be the one species, other than my own, that I've spent the most time sitting in cafés with. Our species shares a ten-thousand-year history with house sparrows. About the same length of history that we share with cows. But while cows have been deliberately molded by humans for our own use, sparrows have retained their own dominion. They live with us, not domesticated by us. The house sparrow has become a totally new version of itself that fits an urban environment. Just as lichens, linden trees, and songbirds have done. It's as if the tough urban landscape forces the city's living creatures to display their vitality.

IS COMPARING AN Antarctic penguin colony with a flock of inner-city sparrows going a bit too far? Is there wilderness of any kind in the city? It depends on how you use the word. The dictionary definition states that wilderness is a "natural area that is not significantly disturbed by human activity," which would mean no wilderness exists in the city. But the term can also be defined more subjectively. The English word for wild is derived from the Proto-Germanic word *wildia*, which means "uncultivated, untamed, uncontrolled." The word "wild" can be defined as "a wild, free, or natural state or existence."

Even though sparrows, ants, and lichens are affected by humans and the influence we exert, they still live by their own rules. Many of the city's animals have changed—and probably genetically too—in the face of humans and the landscapes we have created, but these changes have been beyond our control. These creatures have simply adapted to a new environment. Anyone who has encountered a crow, or an ant, knows that we are not masters of these animals. Turn your back and the moss will grow from the cracks in the concrete. Beneath our well-kept lawns, the world's organisms are alive and ready to take back their undomesticated spaces. After a year of going round the inner city with my loupe and binoculars, I have no doubt. The wilderness has its representatives here too.

I REMEMBER A TEACHER I once had, the nature guide Bjørn Tordsson, who talked about how the city was designed for specific purposes, and how the urban land-scape is designed to make us behave in a certain way. In a lecture he said:

> The streets must expedite the flow of traffic according to carefully regulated principles. Stores are designed for profitable trading. A sports facility tells you what is expected of your physical performance. Objects, tools, and living environments all have clear but invisible

built-in "user manuals," guidelines, and expectations, which govern our comings and goings. The material environment is a force that shapes humans in its image.[12]

I thought about what Tordsson had said. The river valley I was walking through wasn't especially picturesque. But unlike the marked trails through the Antarctic penguin colonies, it wasn't directing me. I was free to explore it. The mallards and sparrows are wild animals, uncontrollable, without a user manual. They just exist, without us having any purpose for them, or vice versa. They live, mate, and die, besiege riverbanks, and plunder café tables. The ants crawl around in the voids and clean the streets. It's nice to encounter the unexpected. Something that just happens, like a sneeze.

I LEFT THE SPARROWS in the bush and threw my coffee cup in a nearby garbage bin. Out of the corner of my eye, I noticed that a crow had perched in one of the trees and was watching me curiously. A year earlier I would have felt repulsed by this gray-black creature. But now I was keen on getting a closer look. I fumbled around for my binoculars before realizing that I'd left them at home. I had started feeling naked without them. Two ducks changed course and swam toward the trampled grass on the riverbank where I stood. They hopped ashore and waddled expectantly in my direction, while a third duck came in for landing so close that I could hear the rush of its feathers as it plowed a furrow in the river. Two magpies that had been evaluating the situation flew down from the willow trees and sat at the edge of the feeding area.

In what took only a moment I was surrounded by birds. Beautiful, wild creatures that allowed me right into their lives. Nothing like this had ever happened to me beyond the city limits. Even the penguins in Antarctica, which on land have lived virtually without enemies, don't allow humans to get as close as urban birds do. There are countless examples of animals that grew up without any fear of humans and have been wiped out by them. But cities are the exception. The brave birds survive in the city, not the timid ones. It's in the city that I can get a close look at how they adapt to the new landscapes we build, how they live and die. And all without emitting any carbon dioxide or investing in anything fancier than a pair of binoculars.

WHY HAD I NEVER looked at crows and mallards with the same enthusiasm that I had studied penguins?

Perhaps our idea of spectacular nature, untouched by human hands, has stood in the way. An expectation that nature has value because it is pure and magnificent and distinct from humans. In other words, the same expectation I derided the Antarctic tourists for, as they attempted to photograph penguins without getting any other red-jacketed visitors in the background.

But I had no reason to look down on the tourists. We were all equally detached from the landscape we sailed past. None of us belonged there. None of us had time to get to know the place, other than to marvel at a few curated snapshots.

A LONE COMMON GULL hovered above the trees. It reminded me of the silence at Fleinvær, and the look in Magne Jensen's eyes beneath his black hat.

Magne knew his home islands by heart. He knew when the white wagtails would return in the spring, and where their permanent nesting sites were. He knew where the otters made paths in the grass. He told me that he could physically feel when something wasn't right there, like the silence, the absence of seagull cries, which had come in recent years.

Magne had a different connection to the natural world around him than the one you experience while studying biology. It was a connection I had never felt but that I somehow envied him for. I have moved between cities, neighborhoods, and countries, between landscapes that are changing, but I've never had either the time or the attentiveness to notice it.

It took a voyage to Antarctica, to the ultimate alien landscape, for me to understand how alien I was to the common orange lichen and the seagulls, to the nature around me in what was after all the place I called home.

IN HER BOOK *Rambunctious Garden,* Emma Marris writes that we have lost nature in two senses of the word.[13]

The first is quite specific. We are losing land, species, and habitats. Where there were once forests, trees, rivers, and marshes, there are now houses, roads, wind turbines, and football stadiums. Less than half a percent, measured by weight, of all the animals living on Earth are wild mammals and birds. The world's domestic animals weigh over ten times as much.[14] Wild bird and animal species are decreasing, and there are far fewer of each species.[15]

But we have also lost nature—for ourselves. We've lost sight of it. Our nature experiences come from social media, where they are near perfect. Or from TV. We watch nature filmed from spectacular angles, with a gripping soundtrack to build the excitement. The nature on-screen looks nothing like what we see around us on a daily basis. It compounds this feeling that if something is undramatic and commonplace, unfiltered or unaccompanied by a narrator's voice, it is not proper nature.

The idea of "real" nature having to be untouched and splendid may be contributing to the alienation city dwellers feel toward the heavily urbanized nature that surrounds us every day. It makes it easier for us to view urban nature as worthless, to view the weeds growing through the cracks in the asphalt, the crows, and the yellow lichens as dilapidated props in our daily lives.

THE BUTTERFLY LOVER and biologist Robert Pyle was one of the first to name the gulf between humans and nature that opens up in the city. He called it the extinction of experience: "It is not just about losing the personal benefits of the natural high. It also implies a cycle of disaffection that can have disastrous consequences."[16]

According to Pyle, as the nature around us is destroyed, our connection to it is eroded at the same rate. Many of us cannot tell bird species apart, so we're unaware that the wrens, dunnocks, and yellowhammers are disappearing from our parks, leaving only the great tits behind—and there are declining numbers of those birds too. Since we haven't learned what lichens and insects live among the cracks in urban tree bark, or what kind of animals raise their young behind flaking bark, the fact that the city skyline is losing increasingly more of its large round treetops doesn't feel like such a great loss. And once these treetops are gone, we have missed our chance to find out what it is we have lost.

When you surround yourself with inadequate and impoverished nature, you experience less of what Pyle describes as "the natural high"—the well-documented positive effect nature has on mental and physical health.[17] But that's not all. When we no longer have access to nature, it also erodes our emotional connection to it. Why should we care about creatures we have never seen or heard?

This is the downward spiral that Pyle described: We feel so little affiliation with nature's species and processes that we're incapable of seeing that what's actually there has value. Nor are we able to see that it is changing, or disappearing.[18] This is how we become accustomed to a city where the green areas are gradually being built over and where fewer and fewer birds sing in the parks. We are losing what they give us. And without people who know and care about every little patch of urban nature, who are fond of bats and lichens, the destruction will be able to continue at an even faster rate.

BUT THIS YEAR I have also found that the spiral can go the opposite way.

It's not like I am settled in the inner city. The idea of leaving it for somewhere more abundant in nature crosses my mind regularly. A desire for a sense of belonging might explain why I am always dreaming of moving closer to the mountains—where you can see the starry sky without the dull filter of artificial light—or to the bustling life in the forest.

I have *tried* moving out of town. But it has never turned out well. Life in the country is nice for a while, but I have always come back to the city. Back to the seagulls, bats, and lichens, and the occasional starling.

AFTER I WAITED FRUITLESSLY for a few minutes, the ducks waddled back into the water. The crow had lost interest and flown away. Even the sparrows had fallen silent for the evening. I stood on the riverbank and gazed out at the illuminated city.

This narrow riverside park, with its willow trees and sparrows, is not a forest, I thought. The landscape before me couldn't replace anything other than itself. But that doesn't necessarily mean that urban nature—this mixture of old and new, of refugees, rarities, and globetrotters, of old acquaintances and species that have never lived together before in Earth's history—isn't valuable. Perhaps it's time we got to know it.

Thank You

T HANKS TO EVERYONE I walked with or talked with (or both): Anne Sverdrup-Thygeson, Arild Breistøl, Audun Eriksen, Bjørn Olav Tveit, Charlotte Bjorå, Dag Inge Lønning, Einar Timdal, Erik Joner, Geir Sonerud, Katelyn Solbakk, Kjell Isaksen, Klaus Høiland, Knut-Petter Sætre Langlo, Kristoffer Bøhn, Linda Jolly, Magne Flåten, Magne Jensen, Per Marstad, and Sven Sunde. I have relied heavily on friends and acquaintances, former lecturers, old colleagues, and others for comments and professional input: Anders Often, Anders Thylén, Guren Efferus, Guro Løseth, Jens-Christian Svenning, Jo Skeie Hermansen, Jon Gunnar Jørgensen, Jon Tjøgersen, Saga Svavarsdottir, Sasha Waltå, Sigmund Hågvar, Stephen Mosley, Tore Oldeide Elgvin, Trude Myhre, and Unni Vik. Many thanks to Unni Vik who did a thorough proofreading job. I have also had valuable read-throughs from Anne Gerd Grimsby Harr, Annie Aas Hovind, Eirik Vigerum Jensen, Georg Blichfeldt, Gunnar Mikalsen Kvifte, Helga Eggebø, Kristin Langdalen, Lars Petter Røyset Marthinsen, Markus Brun Hustad, and Silje Kiær Andersen. And thank you to everyone who offered their crow stories on Underskog and in other forums.

THE NORWEGIAN EDITION was published with financial support from the Norwegian Non-Fiction Writers and Translators Association, and the research project Invisible Infrastructures (project number 259888 / E50), supported by the Research Council of Norway.

Appendix

Author's Comments on the Text

On Urbanization and Biodiversity

Although urban nature is in many ways richer than I initially thought, it is not a substitute for less humanly affected nature. Urbanization is one of the most significant causes of the catastrophic loss of biodiversity, and is generally bad news for the world's species, even if there are some glimmers of hope.[1] Since 1992 the cities of the world have doubled in size, and this growth is expected to increase. Old-growth forests, wetlands, and waterfall gorges support species and ecosystems that will never manage to live or thrive in the city. And humans depend more on these species and ecosystems than we might think. Human cities are a patchwork of landscapes with different histories, but many of them are harsh, dry, hot landscapes that very few species can live in.

The growth of our cities has eliminated the local biological diversity that was originally there, and this diversity has been largely replaced by conformity.[2] Many of the world's cities have cliff-like ecosystems that resemble each other more than they do their rural surroundings.[3]

On Gender Balance

Male interviewees predominate in this book. The most significant reason is that species nerds are (at least in Norway) still often men. This reality may also have something to do with the fact that although a higher proportion of women are studying biology, species knowledge and field-work have been sadly deprioritized in Norwegian biology education over recent decades. Nevertheless, several indications suggest that this trend is about to reverse. I hope and believe there will be more female species nerds in the years to come.

Notes

Introduction: An Old Acquaintance

1   The IMBIE team (2018). Mass balance of the Antarctic Ice Sheet from 1992 to 2017. *Nature, 558*: 219–222. https://doi.org/10.1038/s41586-018-0179-y

2   Davies, B. J., Carrivick, J. L., Glasser, N. F., Hambrey, M. J., & Smellie, J. L. (2012). Variable glacier response to atmospheric warming, northern Antarctic Peninsula, 1988–2009. *The Cryosphere, 6*: 1031–1048. https://doi.org/10.5194/tc-6-1031-2012

3   An approximate estimate of diesel consumption during normal sea currents and weather conditions, in an equally large boat, at a cruising speed of 18 knots, is 30,000 liters of diesel per day, or well over 1,000 liters per hour. A thousand liters is equivalent to 265 gallons. Personal statement, Ivan Johansen, engineer. I have contacted several individuals in the cruise company in question, to get a precise estimate, without getting a response.

4   Naveen, R., Lynch, H. J., Forrest, S., Mueller, T., & Polito, M. (2012). First direct, site-wide penguin survey at Deception Island, Antarctica, suggests significant declines in breeding chinstrap penguins. *Polar Biology, 35*(12): 1879–1888. https://doi.org/10.1007/s00300-012-1230-3

5   Other Antarctic penguin colonies are also experiencing a dramatic decline. Source: Sander, M., Balbão, T. C., Costa,

E. S., dos Santos, C. R., & Petry, M. V. (2007). Decline of the breeding population of *Pygoscelis antarctica* and *Pygoscelis adeliae* on Penguin Island, South Shetland, Antarctica. *Polar Biology, 30*: 651–654. https://doi.org/10.1007/s00300-006-0218-2

6 Less sea ice changes the ocean's ecosystems, and this change is reflected above sea level.

7 The wilderness is exclusive, because there is so little left of it. Seventy-five percent of the earth's land area and 65 percent of the sea has changed significantly as a result of direct human impact. Source: Díaz, S., Settele, J., Brondízio, E. S. (co-chairs) et al. (2019). *The global assessment report on biodiversity and ecosystem services: Summary for policymakers.* Intergovernmental Science-Policy Platform on Biodiversity and Ecosystem Services (IPBES). p. 11. https://ipbes.net/sites/default/files/inline/files/ipbes_global_assessment_report_summary_for_policymakers.pdf

January: The Urbanite's Reflection

1 In Trondheim, up to 10,000 crows, including a small number of other corvids, have been recorded at some gathering places. Husby, M. Kråke [Crow]. *Store Norske Leksikon.* Retrieved May 25, 2022, from https://snl.no/kr%c3%a5ke [Norwegian only]

2 Sonerud, G. A., & Fjeld, P. E. (1984). Searching and caching behaviour in hooded crows: An experiment with artificial nests. *Fauna norvegica, 8*: 18–21. https://www.researchgate.net/publication/259292971_Searching_and_caching_behaviour_in_hooded_crows_-_an_experiment_with_artificial_nests

3 Rosati, A. G., Stevens, J. R., Hare, B., & Hauser, M. D. (2007). The evolutionary origins of human patience: Temporal preferences in chimpanzees, bonobos, and human adults. *Current Biology, 17*(19): 1663–1668. https://doi.org/10.1016/j.cub.2007.08.033. A similar test, the classic

marshmallow test, was carried out in children in the 1960s. Those results have recently been questioned, but adults may well act differently from children. See Calarco, J. M. (2018, June 1). Why rich kids are so good at the marshmallow test. *The Atlantic.* https://www.theatlantic.com/family/archive/2018/06/marshmallow-test/561779

4   Marzluff, J., & Angell, T. (2012). *Gifts of the crow: How perception, emotion and thought allow smart birds to behave like humans.* Free Press, a division of Simon & Schuster, Chapter 5.

5   Dally, J., Emery, N., & Clayton, N. (2010). Avian theory of mind and counter espionage by food-caching western scrub-jays (*Aphelocoma californica*). *European Journal of Developmental Psychology, 7*(1): 17–37. https://doi.org/10.1080/17405620802571711

6   Heinrich, B., & Pepper, J. W. (1998). Influence of competitors on caching behaviour in the common raven, *Corvus corax. Animal Behaviour, 56*(5): 1083–1090. https://doi.org/10.1006/anbe.1998.0906

7   Bugnyar, T., & Kotrschal, K. (2004). Leading a conspecific away from food in ravens (*Corvus corax)? Animal Cognition, 7*(2): 69–76. https://doi.org/10.1007/s10071-003-0189-4

8   Emery, N. J., & Clayton, N. S. (2004). The mentality of crows: Convergent evolution of intelligence in corvids and apes. *Science, 306*(5703): 1903–1907. https://doi.org/10.1126/science.1098410

9   Astington, J. W., & Edward, M. J. (2010). The development of theory of mind in early childhood. In Tremblay, R. E., Boivin, M., Peters, R. DeV. (Eds.), Zelazo, P. D. (Topic Ed.). *Encyclopedia on Early Childhood Development* [online]. Retrieved May 25, 2022, from https://www.child-encyclopedia.com/social-cognition/according-experts/development-theory-mind-early-childhood

10 Calero, C. I., Salles, A., Semelman, M., & Sigman, M. (2013). Age and gender dependent development of Theory of Mind in 6- to 8-years old children. *Frontiers in Human Neurosciences, 7*: 281. https://doi.org/10.3389/fnhum.2013.00281

11 Jarvis, E. D., Mirarab, S., Aberer, A. J. et al. (2014). Whole-genome analyses resolve early branches in the tree of life of modern birds. *Science, 346*(6215): 1320–1331. https://doi.org/10.1126/science.1253451

12 Husby, M. Kråke [Crow].

13 Riechelmann, C. (2014). *Kråker: Et portrett*. Forlaget Press.

14 Hansen, H., Smedshaug, C. A., & Sonerud, G. A. (2000). Preroosting behaviour of hooded crows (*Corvus corone cornix*). *Canadian Journal of Zoology, 78*(10): 1813–1821. https://doi.org/10.1139/cjz-78-10-1813

15 Sonerud, G. A., Smedshaug, C. A., & Bråthen, Ø. (2001). Ignorant hooded crows follow knowledgeable roost-mates to food: Support for the information centre hypothesis. *Proceedings of the Royal Society B, 268*(1469): 827–831. https://doi.org/10.1098/rspb.2001.1586

16 Wright, J., Stone, R. E., & Brown, N. (2003). Communal roosts as structured information centres in the raven, *Corvus corax*. *Journal of Animal Ecology, 72*(6): 1003–1014. https://doi.org/10.1046/j.1365-2656.2003.00771.x

17 Marzluff, J., Heinrich, B., & Marzluff, C. S. (1996). Raven roosts are mobile information centres. *Animal Behaviour, 51*(1): 89–103. https://doi.org/10.1006/anbe.1996.0008

18 Wright, J., Stone, R. E., & Brown, N. (2003). Communal roosts as structured information centres in the raven, *Corvus corax*.

19 The name "crow" follows a proud tradition of naming birds after the sound they make. de Caprona, Y. (2013). *Norsk etymologisk ordbok*. Kagge Forlag. [Norwegian only]

20 de Vaan, M. (2008). *Etymological Dictionary of Latin—and the other Italic Languages.* Brill, p. 150.

21 de Vaan, M. (2008). *Etymological Dictionary of Latin—and the other Italic Languages.* p. 153.

22 Chamberlain, D. E., & Cornwell, G. W. (1971). Selected vocalizations of the common crow. *The Auk, 88*(3): 613–634. https://doi.org/10.1093/auk/88.3.613

23 Chamberlain, D. E., & Cornwell, G. W. (1971). Selected vocalizations of the common crow.

24 This study concerns a different member of the crow family than the Norwegian hooded crow. Mates, E. A., Tarter, R. R., Ha, J. C., Clark, A. B., & McGowan, K. J. (2015). Acoustic profiling in a complexly social species, the American crow: Caws encode information on caller sex, identity and behavioural context. *Bioacoustics, 24*(1): 63–80. https://doi.org/10.1080/09524622.2014.933446

25 This study concerns a different member of the crow family than the Norwegian hooded crow. Schalz, S., & Izawa, E. (2020). Language discrimination by large-billed crows. In *The Evolution of Language: Proceedings of the 13th International Conference* (Evolang13): 380–390. https://doi.org/10.17617/2.3190925

26 Marzluff, J., & Angell, T. (2012). *Gifts of the crow*, Chapter 9: Awareness.

27 Marzluff, J., Walls, J., Cornell, H. N., Withey, J. C., & Craig D. (2010). Lasting recognition of threatening people by wild American crows. *Animal Behavior, 79*(3): 699–707. https://doi.org/10.1016/j.anbehav.2009.12.022

28 Cornell, H. N., Marzluff, J., & Pecoraro, S. (2012). Social learning spreads knowledge about dangerous humans among American crows. *Proceedings of the Royal Society B, 279*(1728): 499–508. https://doi.org/10.1098/rspb.2011.0957

29 Nile, A. (2016, April 29). Thousands of crows darken Bothell's sky, attracting scholars. *Heraldnet*. https://www.heraldnet.com/news/thousands-of-crows-darken-bothells-sky-attracting-scholars/

30 Swift, K. N. (2015). Wild American crows use funerals to learn about danger [Master's thesis, University of Washington]. https://digital.lib.washington.edu/researchworks/bitstream/handle/1773/33178/Swift_washington_02500_14237.pdf

31 Swift, K. N., & Marzluff, J. (2015). Wild American crows gather around their dead to learn about danger. *Animal Behaviour, 109*: 187–197. https://doi.org/10.1016/j.anbehav.2015.08.021

32 von Bayern, A. M. P., Danel, S., von Auersperg, A. M. I., Mioduszewska, B., & Kacelnik, A. (2018). Compound tool construction by New Caledonian crows. *Scientific Reports, 8*(15676). https://doi.org/10.1038/s41598-018-33458-z

33 Institutt for biovitenskap, Universitetet i Oslo. *Nervesystem*. Retrieved May 26, 2022, from https://www.mn.uio.no/ibv/tjenester/kunnskap/plantefys/leksikon/n/nervesystem.html [Norwegian only]

34 Wang, Y., Brzozowska-Prechtl, A., & Karten, H. J. (2010). Laminar and columnar auditory cortex in avian brain. PNAS, *107*(28): 12676–12681. https://doi.org/10.1073/pnas.1006645107

35 Marzluff, J., & Angell, T. (2012). *Gifts of the crow*, Chapter 2: Birdbrains nevermore.

36 Marzluff, J., & Angell, T. (2012). *Gifts of the crow*, p. 23.

37 Watanabe, S., Sakamoto, J., & Wakita, M. (1995). Pigeons' discrimination of paintings by Monet and Picasso. *Journal of the Experimental Analysis of Behaviour, 63*(2): 165–174. https://doi.org/10.1901/jeab.1995.63-165

38 Roth, G., & Dicke, U. (2005). Evolution of the brain and intelligence. *Trends in Cognitive Sciences, 9*(5): 250–257. https://doi.org/10.1016/j.tics.2005.03.005

39 Lorenz discovered, among other things, that geese bond with whomever they are close to for between 12 and 16 hours after hatching. Even if this is someone other than their own mother, the goslings will start to follow them (for example, Lorenz himself). There are numerous photos of Lorenz walking, and even swimming, with "his" goslings trailing behind. Marzluff, J., & Angell, T. (2012). *Gifts of the crow*, p. 24.

40 Marzluff, J., & Angell, T. (2012). *Gifts of the crow*, p. 23.

41 Morell, V. (2018, February). Think 'birdbrain' is an insult? Think again. *National Geographic.* https://www.nationalgeographic. com/magazine/2018/02/bird-brains-crows-cockatoos-songbirds-corvids/

42 Emery, N. J., & Clayton, N. S. (2004). The mentality of crows.

43 For a thorough discussion of how evolution has worked with mammals and birds' brains, I can recommend Marzluff, J., & Angell, T. (2012). *Gifts of the crow*, Chapter 2: Birdbrains nevermore.

44 Seed, A., Emery, N., & Clayton, N. (2009). Intelligence in corvids and apes: A case of convergent evolution? *Ethology, 115*(5): 401–420. https://doi.org/10.1111/j.1439-0310 .2009.01644.x

45 Sonerud, G. A., & Fjeld, P. E. (1987). Long-term memory in egg predators: An experiment with a hooded crow. *Ornis Scandinavica, 18*(4): 323–325. https://doi.org/10.2307/ 3676904

46 Cnotka, J., Güntürkün, O., Rehkämper, G., Gray, R. D., & Hunt, G. R. (2008). Extraordinary large brains in tool-using

New Caledonian crows (*Corvus moneduloides*). *Neuroscience Letters, 433*(3): 241–245. https://doi.org/10.1016/j.neulet.2008.01.026

47 Mehlhorn, J., Hunt, G. R., Gray, R. D., Rehkämper, G., & Güntürkün, O. (2010). Tool-making New Caledonian crows have large associative brain areas. *Brain Behavioral Evolution, 75*(1): 63–70. https://doi.org/10.1159/000295151

48 Emery, N. J., & Clayton, N. S. (2004). The mentality of crows.

49 Olkowicz, S., Kocourek, M., Lučan, R. K., Porteš, M., Fitch, W. T., Herculano-Houzel, S., & Němec, P. (2016). Birds have primate-like numbers of neurons in the forebrain. *PNAS, 113*(26): 7255–7260. https://doi.org/10.1073/pnas.1517131113

50 Colding, J., Gren, Å., & Barthel, S. (2020). The incremental demise of urban green spaces. *Land, 9*(5): 162. https://doi.org/10.3390/land9050162.

51 Marzluff, J., Miyaoka, R., Minoshima, S., & Cross, D. J. (2012). Brain imaging reveals neuronal circuitry underlying the crow's perception of human faces. *PNAS, 109*(39): 15912–15917. https://doi.org/10.1073/pnas.1206109109

52 Hellesøy, C. (2014, February 21). Kråker er smarte og langsinte [Crows are smart and long-suffering]. *Aftenposten*. Retrieved May 26, 2022, from https://www.aftenposten.no/norge/i/JIGEP/kraaker-er-smarte-oglangsinte [Norwegian only]

March: The Night Singer

1 Samstag, S. O. (2009, April 29). Markavennene på fugletur [The field friends on a bird walk]. *Østlandets blad*. Retrieved May 26, 2022, from https://www.oblad.no/badebyen/markavennene-pa-fugletur/s/2-2.2610-1.3634866 [Norwegian only]

2   Constantine, M. (2013). *The sound approach to birding: A guide to understanding bird sound.* The Sound Approach, p. 124.

3   Constantine, M. (2013). *The sound approach to birding,* Chapter 7: Magnus Robb and "The Blackcaps."

4   Charny, B. (2002, January 2). Birds sing a new tune in wireless era. CNET. Retrieved May 26, 2022, from https://www .cnet.com/news/birds-sing-a-new-tune-in-wireless-era/

5   Constantine, M. (2013). *The sound approach to birding,* p. 93.

6   Constantine, M. (2013). *The sound approach to birding,* p. 109.

7   Odom, K. J., & Benedict L. (2018) A call to document female bird songs: applications for diverse fields. *The Auk, 135*(2): 314–325. https://doi.org/10.1642/auk-17-183.1

8   Webb, W., Brunton, D., Aguirre, J., Thomas, D., Valcu, M., & Dale, J. (2016). Female song occurs in songbirds with more elaborate female coloration and reduced sexual dichromatism. *Frontiers in Ecology and Evolution, 4*(40): 1–8. https://doi.org/10.3389/fevo.2016.00022

9   Constantine, M. (2013). *The sound approach to birding,* p. 88.

10 European Environment Agency (2019). *Environmental Noise in Europe—2020.* EEA report 22/2019. ISSN 1977–8449, p. 15. https://www.eea.europa.eu/publications/ environmental-noise-in-europe

11 Albert, D. G., & Decato, S. N. (2010). The pulse of a city: Noise measurements in Baltimore. In *Acoustical Society of America 159th Meeting Lay Language Papers.* https://acoustics .org/pressroom/httpdocs/159th/albert.htm

12 Brumm, H. (2004). The impact of environmental noise on song amplitude in a territorial bird. *Journal of Animal Ecology, 73*(3): 434–440. https://doi.org/10.1111/j.0021-8790.2004.00814.x

13 Slabbekoorn, H., & Peet, M. (2003). Birds sing at a higher pitch in urban noise. *Nature, 424*(6946), 267. https://doi .org/10.1038/424267a

14 Dowling, J. L., Luther, D. A., & Marra, P. P. (2011). Comparative effects of urban development and anthropogenic noise on bird songs. *Behavioral Ecology, 23*(1): 201–209. https://doi.org/10.1093/beheco/arr176.

15 Partan, S. R., Fulmer, A. G., Gounard, M. A. M., & Redmond, J. E. (2019). Multimodal alarm behavior in urban and rural gray squirrels studied by means of observation and a mechanical robot. *Current Zoology, 56*(3): 313–326. https://doi.org/10.1093/czoolo/56.3.313

16 Reichard, D. G., Atwell, J. W., Pandit, M. M., Cardoso, G. C., Price, T. D., & Ketterson, E. D. (2019). Urban birdsongs: Higher minimum song frequency of an urban colonist persists in a common garden experiment. *Animal Behaviour 170*: 33–41. https://doi.org/10.1016/ j.anbehav.2020.10.007

17 Several examples of how city life can change the incoming species genetically can be found in the highly enjoyable book *Darwin Comes to Town* by Menno Schilthuizen. Schilthuizen, M. (2018). *Darwin comes to town: How the urban jungle drives evolution.* Picador. Chapter 18: *Turdus urbanicus.*

18 The city's warm climate and abundance of lawns, berries, and fruit trees sustain a large population of blackbirds throughout the year. City birds live close together, far more densely than in the forests they originally came from.

19 Humid air carries sound better than dry air. Birds know this, and make the most of it after a rainstorm.

20 Nordt, A., & Klenke, R. (2013). Sleepless in town—drivers of the temporal shift in dawn song in urban European blackbirds. *PLOS ONE, 8*(8): e71476. https://doi.org/ 10.1371/journal.pone.0071476

21 Menno Schilthuizen argues that the tendency to be awake
    at night is one of the characteristics that distinguishes city
    blackbirds from their rural cousins in two separate species.
    Schilthuizen, M. (2018). *Darwin comes to town*, Chapter 18.

April: The War on Ants

1   The approximate (and conservatively) estimated
    number of ants on the planet is ten thousand trillion, that
    is, 10,000 × 10^12, or 10,000,000,000,000,000. Hölldobler,
    B., & Wilson E. O. (1994). *Journey to the ants: A story of
    scientific exploration* (2nd ed.). Harvard University Press.
    Chapter: The dominance of ants. By comparison, as of
    June 2021 we humans are 7.8 billion, or 7,800,000,000.

2   Hölldobler, B., & Wilson E. O. (1994). *Journey to the
    ants*, Chapter: The dominance of ants. This claim has
    been criticized: Moore, H. (2014, September 22). Are all
    the ants as heavy as all the humans? BBC *News*. Retrieved
    on May 26, 2022, from https://www.bbc.com/news/
    magazine-29281253.

3   There are some exceptions: above the tree line, for exam-
    ple, you will find very few ants.

4   Hölldobler, B., & Wilson E. O. (1994). *Journey to the ants*.

5   In humans, several conditions can change external gender
    expression, such as androgen insensitivity syndrome and
    adrenogenital syndrome. Androgen insensitivity syndrome.
    MedlinePlus. Retrieved on May 30, 2022, from https://doi
    .org/10.1007/s00300-006-0218-2

6   Hölldobler, B., & Wilson E. O. (1994). *Journey to the ants*,
    p. 37.

7   Hölldobler, B., & Wilson E. O. (1994). *Journey to the ants*,
    p. 35.

8   Goulson, D. (2019). *The garden jungle, or gardening to save
    the planet*. Jonathan Cape.

9   Yong, E. (2009, September 26). Ants herd aphids with tranquilisers in their footsteps. *National Geographic.* Retrieved on May 30, 2022, from https://www.national geographic.com/science/article/ants-herd-aphids-with-tranquilisers-in-their-footsteps

10 Phillips S. M., & Van Loon, L. J. C. (2011). Dietary protein for athletes: From requirements to optimum adaptation. *Journal of Sports Sciences 29 Suppl 1*: s29–38. https://doi.org/10.1080/02640414.2011.619204

11 Yong, E. (2009, September 26). Ants herd aphids with tranquilisers in their footsteps.

12 Hölldobler, B., & Wilson E. O. (1994). *Journey to the ants.*

13 Völkl, W., Woodring, J., Fischer, M., Lorenz, M. W., & Hoffmann, K. H. (1999). Ant-aphid mutualisms: The impact of honeydew production and honeydew sugar composition on ant preferences. *Oecologia, 118*(4): 483–491. https://doi.org/10.1007/s004420050751

14 Goulson, D. (2015). *A buzz in the meadow.* Picador.

15 Apparently, 99 percent of all organisms communicate primarily through chemical substances. Hölldobler, B., & Wilson E. O. (1994). *Journey to the ants,* Chapter 4: How ants communicate.

16 d'Ettorre P. (2016). Genomic and brain expansion provide ants with refined sense of smell. pnas, *113*(49): 13947–13949. https://doi.org/10.1073/pnas.1617405113

17 Humans have many olfactory cells, but they are remnants of our evolutionary past; we have left most of our ability to perceive reality up to our eyes and ears. Bushdid, C., Magnasco, M. O., Vosshall, L. B., & Keller, A. (2014). Humans can discriminate more than 1 trillion olfactory stimuli. *Science, 343*(6177): 1370–1372. https://doi.org/10.1126/science.1249168

18 Stoyanov, G. S., Matev, B. K., Valchanov, P., Sapundzhiev, N., & Young, J. R. (2018). The human vomeronasal (Jacobson's) organ: A short review of current conceptions, with an English translation of Potiquet's original text. *Cureus,* 10(5): e2643. https://doi.org/10.7759/cureus.2643

19 Hölldobler, B., & Wilson E. O. (1994). *Journey to the ants,* Chapter: How ants communicate.

20 Hölldobler, B., & Wilson E. O. (1994). *Journey to the ants,* Chapter: How ants communicate.

21 Hölldobler, B., & Wilson E. O. (1994). *Journey to the ants,* Chapter: The origin of cooperation.

22 d'Ettorre P. (2016). Genomic and brain expansion provide ants with refined sense of smell.

23 Hallmann, C. A., Sorg, M., Jongejans, E., Siepel, H., Hofland, N., Schwan, H., Stenmans, W., Müller, A., Sumser, H., Hörren, T., Goulson, D., & de Kroon, H. (2017). More than 75 percent decline over 27 years in total flying insect biomass in protected areas. *PLOS ONE 12*: e0185809. https://doi.org/10.1371/journal.pone.0185809

24 Cherry R. (2015). Insects and divine interventions. *American Entomologist, 61(2)*: 81–84. https://doi.org/10.1093/ae/tmv001

25 Cherry R. (2015). Insects and divine interventions.

26 Russell, E. P., III. (1996). "Speaking of annihilation": Mobilizing for war against human and insect enemies, 1914–1945. *The Journal of American History, 82(4)*: 1505–1529. https://doi.org/10.2307/2945309

27 Russell, E.P., III. (1996). "Speaking of annihilation."

28 Mata, L., Threlfall, C. G., Williams, N. S. G., Hahs, A. K., Malipatil, M., Stork, N. E., & Livesley, S. J. (2017). Conserving herbivorous and predatory insects in urban green spaces. *Scientific Reports, 7*. https://doi.org/10.1038/srep40970

29 Baldock, K. C. R., Goddard, M. A., Hicks, D. M., Kunin, W. E., Mitschunas, N., Osgathorpe, L. M., Potts, S. G., Robertson, K. M., Scott, A. V., Stone, G. N., Vaughan, I. P., & Memmott, J. (2015). Where is the UK pollinator biodiversity? The importance of urban areas for flower-visiting insects. *Proceedings of the Royal Society B, 282*(1803): 20142849. https://doi. org/10.1098/rspb.2014.2849

30 Hall, D. M., Camilo, G. R., Tonietto, R. K., Ollerton, J., Ahrné, K., Arduser, M., Ascher, J. S., Baldock, K. C., Fowler, R., Frankie, G., Goulson, D., Gunnarsson, B. et al. (2017). The city as a refuge for insect pollinators. *Conservation Biology, 31*(1): 24–29. https://doi. org/10.1111/cobi.12840

June: The Seagull Paradox

1 It is forbidden for anyone other than landowners to collect eggs in Norway. There are strict restrictions on which species can be collected and when it can be done. Ministry of Climate and the Environment (2017). *Forskrift om Jakt og fangsttider m.m. 1. april 2017–31. mars 2022.* https://lovdata .no/dokument/SFO/forskrift/2017-01-25-106 [Norwegian only]

2 King, A. (2021, June 15). Why seagulls are making their homes in our cities. *BBC News.* Retrieved on May 26, 2022, from https://www.bbc.com/future/article/20210615-why-sea-gulls-are-making-their-homes-in-our-cities

3 Helberg, M. Gråmåke [Gray gull]. *Store Norske Leksikon.* Retrieved May 26, 2022, from https://snl.no/gr%c3%a5m %c3%a5ke [Norwegian only]

4 Helberg, M. Fiskemåke [Herring gull]. *Store Norske Leksikon.* Retrieved May 26, 2022, from https://snl.no/fiskem%c3 %a5ke [Norwegian only]

5 Artsdatabanken. *Larus canus Linnaeus, 1758.* Retrieved May 26, 2022, from https://artsdatabanken.no/rodliste 2015/rodliste2015/norge/3681 [Norwegian only]

6 Herring gulls are seeing a 50 percent decline in their
population south of Stadt. *Rødlistevurdering fiskemåke*. [Nor-
wegian only]

7 King, A. (2021, June 15). Why seagulls are making their
homes in our cities.

8 Iversen, N. (2021, August 27). Mink. *Visitor Centre Carnivore*.
Retrieved on May 26, 2022, from https://rovdyrsenter.no/
facts-about-large-carnivores/about-the-marten-
family-mustelids/mink/

9 The science building is home to the largest seagull colony
in the municipality, with 78 breeding pairs as of 2019. Arild
Breistøl, personal statement.

10 King, A. (2021, June 15). Why seagulls are making their
homes in our cities.

11 *Columba*. Online Etymology Dictionary. Retrieved on
June 1, 2022, from https://www.etymonline.com/
search?q=columba

12 Personal statement, Magne Flåten.

13 Stein, M. (2019, March 11). Pigeon milk is a nutritious treat
for chicks. *Audubon*. Retrieved on May 26, 2022, from
https://www.audubon.org/news/pigeon-milk-nutritious-
treat-chicks

14 Natural History Museum (2019, June 6). Fuglearter
observer i Botanisk hage [Bird species observed in botanical garden]. Naturhistorisk museum, Universitetet i Oslo.
Source https://www.nhm.uio.no/kunnskapsunivers/
zoologi/fugl/fugler-i-hagen/ [Norwegian only]

15 Fuller, R. A., Tratalos, J., & Gaston, K. J. (2009). How
many birds are there in a city of half a million people?
Diversity and Distributions, *15*(2): 328–337. https://doi
.org/10.1111/j.1472-4642.2008.00537.x

16 Fuller, R. A., Tratalos, J., & Gaston, K. J. (2009). How
many birds are there in a city of half a million people?

17 Schiltzhuisen, M. (2018). *Darwin comes to town: How the urban jungle drives evolution.* Picador.

18 Rosenberg, K. V., Dokter, A. M., Blancher, P. J., Sauer, J. R., Smith, A. C., Smith, P. A., Stanton, J. C., Panjabi, A., Helft, L., Parr, M., & Marra, P. P. (2019). Decline of the North American avifauna. *Science, 366*(6461): 120–124. http://doi.org/10.1126/science.aaw1313

19 Inger, R., Gregory, R., Duffy, J. P., Stott, I., Voříšek, P., & Gaston, K. J. (2015). Common European birds are declining rapidly while less abundant species' numbers are rising. *Ecology Letters, 18*(1): 28–36. https://doi.org/10.1111/ele.12387

20 McMahon, B. J., Doyle, S., Gray, A., Kelly, S. B. A., & Redpath, S. M. (2020). European bird declines: Do we need to rethink approaches to the management of abundant generalist predators? *Journal of Applied Ecology, 57*(10): 1885–1890. https://doi.org/10.1111/1365-2664.13695

21 Rosenberg, K. V., Dokter, A. M., Blancher, P. J. et al. (2019). Decline of the North American avifauna.

22 Nettavisen (2018, January 26). Vi ser en tilbakegang for kjøttmeisen som folk flest kanskje ikke tenker over [We are seeing a decline in great tit numbers which most people are perhaps unaware of]. *Nettavisen Nyheter.* Retrieved on May 26, 2022, fromhttps://www.nettavisen.no/nyheter/innenriks/vi-ser-en-tilbakegang-for-kjottmeisen-som-folk-flestkanskje-ikke-tenker-over/s/12-95-3423409833 [Norwegian only]

23 Haupt, L. L. (2009). *Crow planet: Finding our place in the Zoopolis.* Little, Brown and Company.

July: The Ghosts of the City

1   This metaphor is inspired by "octopus lindens," a description for old lindens, in Brandrud, T. E., Hanssen,

O., Sverdrup-Thygeson, A., & Ødegaard, F. (2011). *Kalklindeskog—et hotspot-habitat: Sluttrapport under ARKO-prosjektets periode II.* NINA report 711. Retrieved on May 26, 2022, from https://www.nina.no/archive/nina/PppBasePdf/rapport/2011/711.pdf [Norwegian only]

2 Bond, W. J., & Midgley, J. J. (2001). Ecology of sprouting in woody plants: the persistence niche. *Trends in Ecology and Evolution, 16*(1): 45–51. https://doi.org/10.1016/s0169-5347(00)02033-4

3 Brandrud, T. E., Hanssen, O., Sverdrup-Thygeson, A., & Ødegaard, F. (2011). *Kalklindeskog—et hotspot-habitat,* p. 14. [Norwegian only]

4 Nedkvitne, K. (1997). *Lind i norsk natur og tradisjon.* Norsk skogbruksmuseum, særpublikasjon 12, p. 81. [Norwegian only]

5 Høeg, O. A. (1974). *Planter og tradisjon.* Norges almenvitenskapelige forskningsråd, pp. 633–640. [Norwegian only]

6 Sandberg, S. (2019). *Treboka: Fakta og fortellinger om norske trær.* Gyldendal Norsk Forlag, p. 263. [Norwegian only]

7 Anne Sverdrup-Thygeson's text on the same topic was published in the newspaper *Klassekampen.* Sverdrup-Thygeson, A. (2018, July 13). Plen, plenere, plenty. *Klassekampen.* https://klassekampen.no/utgave/2018-07-13/plenplenereplenty [Norwegian only]

8 Briske., D. D. (1996). Strategies of plant survival in grazed systems: A functional interpretation. In Hodgson, J., & Illius, A. W. (Eds.) *The ecology and management of grazing systems,* pp. 37–67. CAB International.

9 Tsing, A. L., Swanson, H. A., Gan, E., & Bubandt, N. (Eds.). (2017). *Arts of living on a damaged planet: Ghosts and monsters of the Anthropocene.* University of Minnesota Press, p. 6.

10 Shoshani, J., Goren-Inbar, N., & Rabinovich, R. (2001). A stylohyoideum of *Palaeoloxodon antiquus* from Gesher Benot Ya'aqov, Israel: morphology and functional inferences. The World of Elephants 1st International Congress, Rome, 2001: 665–667.

11 Stuart, A. J. (2005). The extinction of woolly mammoth (*Mammuthus primigenius*) and straight-tusked elephant (*Palaeoloxodon antiquus*) in Europe. *Quaternary International*, 126–128:171–177. https://doi.org/10.1016/j.quaint.2004.04.021

12 Pigott, D. (2012). *Lime-trees and basswoods (A biological monograph of the genus* Tilia). Cambridge University Press.

13 Bond, W. J., & Midgley, J. J. (2001). Ecology of sprouting in woody plants.

August: Stories From the Underground

1 The final compost soil is about 20 percent of the original mass. Personal statement, Linda Jolly.

2 Montgomery, D. R. (2007). *Dirt: The erosion of civilizations*. University of California Press.

3 Diamond, J. (2005). *Collapse: How societies choose to fail or succeed*. Viking Press.

4 Guilland, C., Maron, P. A., Damas, O., & Ranjard, L. (2018). Biodiversity of urban soils for sustainable cities. *Environmental Chemistry Letters, 16*: 1267–1282. https://doi.org/10.1007/s10311-018-0751-6

5 Rasse, D., Økland, I. H., Bárcena, T. G., Riley, H., Martinsen, V., Sturite, I., Joner, E., O'Toole, A., Øpstad, S., Cottis, T., & Budai, A. E. (2019). *Muligheter og utfordringer for økt karbonbinding i jordbruksjord [Opportunities and challenges for increased carbon sequestration in agricultural soil]*. NIBIO 5/36, p. 14. https://nibio.brage.unit.no/nibio-xmlui/bitstream/handle/11250/2591077/NIBIO_RAPPORT_2019_5_36.pdf [Norwegian only]

6 Personal statement, Erik Joner.

7 Pommeresche, R. (2018, February 6). *Biologisk jordstruktur*. Agropub. Retrieved on May 26, 2022, from https://www.agropub.no/fagartikler/biologisk-jordstruktur [Norwegian only]

8 Several factors are at play during the formation of humus, but they are too extensive to include in this context.

9 Wictionary. *Homo*. Retrieved on May 26, 2022, from https://en.wiktionary.org/wiki/homo#Latin

10 Rasse, D., Økland, I. H., Bárcena, T. G. et al. (2019). *Muligheter og utfordringer for økt karbonbinding i jordbruksjord*, p. 14. [Norwegian only]

11 Strozyk, J. (2017). The laboratory study of shear strength of the overconsolidated and quasi-overconsolidated fine-grained soil. *IOP Conference Series: Earth and Environmental Science, 95*(2): 022055. https://doi.org/10.1088/1755-1315/95/2/022055

12 Mellilo, J., & Gribkoff, E. (2021, April 15). Soil-based carbon sequestration. *MIT Climate Portal*. Retrieved on May 26, 2022, from https://climate.mit.edu/explainers/soil-based-carbon-sequestration

13 Stewart, A. (2004). Darwin's worms. *The Wilson Quarterly Archives*. Retrieved on May 26, 2022, from http://archive.wilsonquarterly.com/essays/darwins-worms

14 *Casting about: Darwin on worms*. Darwin Correspondence Project. University of Cambridge. Retrieved on May 26, 2022, fromhttps://www.darwinproject.ac.uk/commentary/life-sciences/casting-about-darwin-worms

15 "Judging by their eagerness for certain kinds of food, they must enjoy the pleasures of eating." Darwin, C. (1881). *The formation of vegetable mould through the action of worms, with observation of their habits*. John Murray, p. 34.

16 Fløistad, E. (2018, September 9). Nyttige nematoder [Use-
ful nematodes]. NIBIO. Retrieved on May 26, 2022, from
https://www.nibio.no/nyheter/nyttige-nematoder [Nor-
wegian only]

17 Rukke, B. A. (2006, 29.11). Spretthaler [Springtails]. Folke-
helseinstitutett (FHI). Retrieved on May 26, 2022, from
https://www.fhi.no/nettpub/skadedyrveilederen/
smadyr-andre/spretthaler/ [Norwegian only]

18 Wired. (2016, June 20). *Behold the super weird face-to-face sex
of the springtail* [video]. YouTube. https://www.youtube
.com/watch?v=uxRNthmhjFW

19 Honey mushrooms are not fungi; they are decomposing
mushrooms. Spilde, I. (2003, April 2). Det største i verden
[The largest in the world]. Forskning.no. Retrieved on
May 26, 2022, from https://forskning.no/sopp-skog-plant-
everden/det-storste-i-verden/1064307 [Norwegian only]

20 Hoorman, J. J., & Islam, R. (2010, September 7). Under-
standing soil microbes and nutrient recycling [SAG-16 fact
sheet]. College of Food, Agricultural, and Environmental
Sciences, Ohio State University Extension. Retrieved on
May 26, 2022, from https://ohioline.osu.edu/factsheet/
SAG-16

21 Clemmensen, K. E., Bahr, A., Ovaskainen, O., Dahlberg,
A., Ekblad, A., Wallander, H., Stenlid, J., Finlay, R. D.,
Wardle, D. A., & Lindahl, B. D. (2013). Roots and associated fungi drive long-term carbon sequestration in boreal
forest. *Science, 339*(6127): 1615–1618. https://doi.org/
10.1126/science.1231923

22 Leaving soil undug can help improve soil structure
through, among other things, improving the soil's ability
to form microaggregates. Ogle, S. M., Alsaker, C., Baldock, J., Bernoux, M., Breidt, F. J., McConkey, B., Regina,
K., & Vazquez-Amabile, G. G. (2019). Climate and soil

characteristics determine where no-till management can store carbon in soils and mitigate greenhouse gas emissions. *Scientific Reports* 9: 11665. https://doi.org/10.1038/s41598-019-47861-7

23 Pommeresche, R., & Seniczak, A. (2018). Jordlevende midd—jordas glemte nyttedyr [fact sheet 7]. *Norsøk*. Retrieved on May 26, 2022, from https://orgprints .org/id/eprint/34140/1/Jordlevende%20midd%20 NORS%C3%98K%20FAGINFO%207%202018%20m%20 ISBN.pdf [Norwegian only]

24 Rasse, D., Økland, I. H., Bárcena, T. G. et al. (2019). *Muligheter og utfordringer for økt karbonbinding i jordbruksjord*, p. 14. [Norwegian only]

25 Rasse, D., Økland, I. H., Bárcena, T. G. et al. (2019). *Muligheter og utfordringer for økt karbonbinding i jordbruksjord*, p. 14.

26 Rasse, D., Økland, I. H., Bárcena, T. G. et al. (2019). *Muligheter og utfordringer for økt karbonbinding i jordbruksjord*, p. 12.

27 Rasse, D., Økland, I. H., Bárcena, T. G. et al. (2019). *Muligheter og utfordringer for økt karbonbinding i jordbruksjord*, p. 5.

28 Rasse, D., Økland, I. H., Bárcena, T. G. et al. (2019). *Muligheter og utfordringer for økt karbonbinding i jordbruksjord*, p. 8.

29 There are several types of root fungi; in this chapter I only write about ectomycorrhizae.

30 Watkinson, S. C., Boddy, L., & Money N. P. (2015). *The Fungi*. Elsevier. Chapter: Mycorrhiza.

31 Reber, S. O., Siebler, P. H., Donner, N. C., Morton, J. T., Smith, D. G., Kopelman, J. M., Lowe, K. R., Wheeler, K. J.,

Fox, J. H., Hassell, J. E., Greenwood, B. N., & Jansch, C. et al. (2016). Immunization with a heat-killed preparation of the environmental bacterium *Mycobacterium vaccae* promotes stress resilience in mice. PNAS, *113*(22): E3130–3139. https://doi.org/10.1073/pnas.1600324113.

32 Roslund, M. I., Puhakka, R., Grönroos, M., Nurminen, N., Oikarinen, S., Gazali, A. M., Cinek, O., Kramná, L., Siter, N., Vari, H. K., Soininen, L., & Parajuli, A. et al. (2020). Biodiversity intervention enhances immune regulation and health-associated commensal microbiota among daycare children. *Science Advances, 6*(42): eaba2578. https://doi.org/10.1126/sciadv.aba2578

October: Marvels of the Darkness

1 Fenton, M. B, Portfors, C. V, Rautenbach, I. L., & Waterman, J. M. (1998). Compromises: Sound frequencies used in echolocation by aerial-feeding bats. *Canadian Journal of Zoology, 76*(6). https://doi.org/10.1139/z98-043

2 Burgin, C. J., Colella, J. P., Kahn, P. L., & Upham, N. S. (2018). How many species of mammals are there? *Journal of Mammalogy, 99*(1): 1–14. https://doi.org/10.1093/jmammal/gyx147. There are over 1,400 species of bats.

3 In other parts of the world, such as parts of Asia, bats are often considered lucky animals. Magne Flåten, personal statement.

4 Laird, T. (2018). *Bats.* Reaktion Books.

5 Wacharapluesadee, S., Tan, C. W., Maneeorn, P., Duengkae, P., Zhu, F., Joyjinda, Y., Kaewpom, T., Chia, W. N., Ampoot, W., Lim, B. L., Worachotsueptrakun, K., & Chen, V. C.-W. et al. (2021). Evidence for SARS-COV-2 related coronaviruses circulating in bats and pangolins in Southeast Asia. *Nature Communications, 12*(1): 972. https://doi.org/10.1038/s41467-021-21240-1

6   Briggs, H. (2020, October 13). Covid: Why bats are
    not to blame, say scientists. *BBC News*. Retrieved on
    May 26, 2022, from https://www.bbc.com/news/
    science-environment-54246473

7   Puig-Montserrat, X., Flaquer, C., Gómez-Aguilera, N.,
    Burgas, A., Mas, M., Tuneu, C., Marquès, E., & López-
    Baucells, A. (2020). Bats actively prey on mosquitoes and
    other deleterious insects in rice paddies: Potential impact
    on human health and agriculture. *Pest Management Science*,
    76(11): 3759–3769. https://doi.org/10.1002/ps.5925

8   Magne Flåten, personal statement.

9   Laird, T. (2018). *Bats*.

10 Ackerman, D. (1992). *The moon by whale light: And other
    adventures among bats, penguins, crocodilians, and whales*.
    Vintage Books. Chapter: In praise of bats.

11 This applies to the echolocation calls the bats use to form
    a picture of the world around them. The social calls they
    use when communicating with each other have a lower
    frequency.

12 Laird, T. (2018). *Bats*.

13 Poptech. (2011). *Shorts: Daniel Kish's echolocation in
    action* [video]. YouTube. https://www.youtube.com/
    watch?v=xaTiyq3uzM4

14 Sjøberg, E., & Blystad, M. H. (2018, June 4). Å gi
    menneskelige egenskaper til dyr kan være problematisk
    [Giving human characteristics to animals can be problem-
    atic.] *Aftenposten*. https://www.aftenposten.no/viten/i/
    zLy18o/aa-gi-menneskelige-egenskaper-til-dyr-kan-vaere-
    problematisk [Norwegian only]

15 Marzluff, J. M., Miyaoka, R., Minoshima S., & Cross, D. J.
    (2012). Brain imaging reveals neuronal circuitry underlying
    the crow's perception of human faces. *PNAS*, 109(39): 15912–
    15917. https://doi.org/10.1073/pnas.1206109109

16 Hammerson, G. A., Kling, M., Harkness, M., Ormes, M., & Young, B. E. (2017). Strong geographic and temporal patterns in conservation status of North American bats, *Biological Conservation, 212*(A): 144–152. https://doi .org/10.1016/j.biocon.2017.05.025

17 Rowse, E. G., Lewanzik, D., Stone, E. L., Harris, S., & Jones, G. (2015). Dark matters: The effects of artificial lighting on bats. In Voigt, C. C., & Kingston, T. (Eds.). *Bats in the anthropocene: Conservation of bats in a changing world.* Springer International Publishing, pp. 187–213. https://doi.org/10.1007/ 978-3-319-25220-9_7

18 Harvey, P. (1980). The advantages of ectothermy for tetrapods. *The American Naturalist, 115*(1): 92–112. https://doi .org/10.1086/283547

19 Frafjord, K. Humleflaggermus [Bumblebee bats]. *Store Norske Leksikon.* Retrieved on May 26, 2022, from https:// snl.no/humleflaggermus [Norwegian only]

20 Schandy, T. Bikolibri [Bee hummingbirds]. *Store Norske Leksikon.* Retrieved on May 26, 2022, from https://snl.no/ bikolibri [Norwegian only]

21 Other researchers believe the number is lower.

22 Ruczyński, I., Ruczyńska, I., & Kasprzyk, K. (2005). Winter mortality rates of bats inhabiting man-made shelters (northern Poland). *Acta Theriologica 50*(2): 161–166. https:// doi.org/10.1007/BF03194479.

23 Eidels, R. R., Whitaker, J. O., Lydy, M. J., & Sparks, D. W. (2013). Screening of insecticides in bats from Indiana. *Proceeding of the Indiana Academy of Science, 121*(2): 133–142.

24 Eidels, R. R., Whitaker, J. O., Lydy, M. J., & Sparks, D. W. (2013). Screening of insecticides in bats from Indiana.

25 We only found one animal. It was either a whiskered or Brandt's bat (it is almost impossible to tell them apart when sleeping). According to bat expert Kristoffer Bøhn, the bat was alive when we found it.

November: The Written Language of the City

1 Draznin, Y. C. (2001). *Victorian London's middle-class housewife*. Greenwood Press, p. 20.

2 Researchers estimate that *Rhizocarpon alpicola* began life in the period after the ice retreated after the last ice age. This corresponds to a growth of 2.65 millimeters per century. Bradley, R. S. (1999). *Paleoclimatology: Reconstructing climates of the Quaternary*. Elsevier, p. 130.

3 Farmer, A. (2013). *Managing environmental pollution*. Routledge, p. 51.

4 Gilbert, O. (2000). *Lichens*. HarperCollins.

5 Lichens have many ways of spreading, both vegetatively and sexually, but common to most is that they spread with the help of wind.

6 Hawksworth, D. L., & Rose, F. (1970). Qualitative scale for estimating sulphur dioxide air pollution in England and Wales using epiphytic lichens. *Nature, 227*(5254): 145–148. https://doi.org/10.1038/227145a0

7 Gilbert, O. (2000). *Lichens*.

8 Einar Timdal referred to Haugsjå, P. K. (1930). Über den einfluss der stadt Oslo auf die flechten vegetation der bäume. *Nyt Magazin for Naturvidenskaberne* 68: 1–116. I have not obtained this article.

9 This is a family of fungi. Lichen gets its Latin name entirely from its fungal partner. The algal partner's name is not represented in the lichen's name.

10 Fortuna, L., & Tretiach, M. (2018). Effects of site-specific climatic conditions on the radial growth of the lichen biomonitor *Xanthoria parietina*. *Environmental Science and Pollution Research, 25*(34): 34017–34026. https://doi.org/10.1007/s11356-018-3155-z

11 de Vera, J.-P., Möhlmann, D., Butina, F., Lorek, A., Wernecke, R., & Ott, S. (2010). Survival potential and photosynthetic activity of lichens under Mars-like conditions: A laboratory study. *Astrobiology, 10*(2): 215–27. https://doi.org/10.1089/ast.2009.0362

12 Asplund, J., & Wardle, D. A. (2017). How lichens impact on terrestrial community and ecosystem properties. *Biological Reviews, 92*(3): 1720–1738. https://doi.org/10.1111/brv.12305

13 The panspermia theory is the idea that life (in the form of lichens, for example) may have spread between planets with meteors, and thus exists in several places in the universe. de la Torre, R., Sancho, L. G., Horneck, G., de los Ríos, A., Wierzchos, J., Olsson-Francis, K., Cockell, C. S., Rettberg, P., Berger, T., de Vera, J.-P. P., Ott, S., Frías, J. M. et al. (2010). Survival of lichens and bacteria exposed to outer space conditions—Results of the Lithopanspermia experiments. *Icarus, 208*(2): 735–748. https://doi.org/10.1016/j.icarus.2010.03.010

December: The Near and the Dear

1 Personal statement, Tore Oldeide Elgvin.

2 Ravinet, M., Elgvin, T. O., Trier, C., Aliabadian, M., Gavrilov, A., & Sætre, G.-P. (2018). Signatures of human-commensalism in the house sparrow genome. *Proceedings of Royal Society B, 285*(1884): 20181246. https://doi.org/10.1098/rspb.2018.1246

3 Ravinet, M., Elgvin, T. O., Trier, C., Aliabadian, M., Gavrilov, A., & Sætre, G.-P. (2018). Signatures of human-commensalism in the house sparrow genome.

4 Symonds, M. R. E., Weston, M. A., van Dongen, W. F. D., Lill, A., Robinson, R. W., & Guay, P.-J. (2016). Time since urbanization but not encephalization is associated with increased tolerance of human proximity in birds. *Frontiers in Ecology and Evolution, 4*: https://doi.org/10.3389/fevo.2016.00117

5 A description of how another city bird, namely the blackbird, has changed personality as a result of its encounters with humans, can be found in the book *Darwin Comes to Town*. Schilthuizen, M. (2018). *Darwin comes to town. How the urban jungle drives evolution*. Picador, Chapter 18: *Turdus urbanicus*.

6 "Companion species." Ravinet, M., Elgvin, T. O., Trier, C., Aliabadian, M., Gavrilov, A., & Sætre, G.-P. (2018). Signatures of human-commensalism in the house sparrow genome.

7 "Eliminate the four pests." Shapiro, J. R. (2001). *Mao's war against nature: Politics and the environment in revolutionary China*. Cambridge University Press, p. 6.

8 Todd, K. (2012). *Sparrow*. pp. 62–63.

9 Shapiro, J. R. (2001). *Mao's war against nature*, p. 88.

10 Platt, J. (2019, July 30). The Great Sparrow Campaign was the start of the greatest mass starvation in history. *Treehugger*. Accessed May 30, 2022. https://www.treehugger.com/the-great-sparrow-campaign-was-the-start-of-the-greatest-mass-4864032

11 Askheim, S., & Garberg, B. Det store spranget [Great Leap Forward]. *Store Norske Leksikon*. Retrieved on May 26, 2022, from https://snl.no/Det_store_spranget

12 Tordsson, B. (2015). Langsommere, dypere, mykere. Inn-
    legg på TUR-konferansen til Den Norske Turistforening.
    http://www.naturliv.no/nalle/bjorn_nalle_tordsson_
    2016.pdf [Norwegian only]

13 Marris, Emma. (2011). *Rambunctious garden: Saving nature in
    a post-wild world.* Bloomsbury, introduction.

14 Bar-On, Y. M., Phillips, R., & Milo, R. (2018). The bio-
    mass distribution on Earth. *PNAS, 115*(25): 6506–6511, fig. 1.
    https://doi.org/10.1073/pnas.1711842115

15 According to the WWF's *Living Planet Report 2020*, the
    world's populations of vertebrates (mammals, birds, rep-
    tiles, amphibians, and fish) have been reduced to an
    average of 30 percent of what they were fifty years ago.
    WWF. (2020). *Living planet report 2020—Bending the curve of
    biodiversity loss.* Almond, R. E. A., Grooten, M., & Petersen,
    T. (Eds.). Gland, Switzerland. https://www.zsl.org/sites/
    default/files/LPR%202020%20Full%20report.pdf

16 "The extinction of experience." Pyle, R. M. (1993). *Thunder
    tree: Lessons from an urban wildland.* Houghton Mifflin.

17 US Department of Agriculture, Forest Service (2018). *Urban
    nature for human health and well-being: A research summary for
    communicating the health benefits of urban trees and green space.*
    (FS–1096). https://www.fs.usda.gov/sites/default/files/
    fs_media/fs_document/urbannatureforhumanhealth
    andwellbeing_508_01_30_18.pdf

18 Soga, M., & Gaston K. J. (2016). Extinction of experience:
    the loss of human–nature interactions. *Frontiers in
    Ecology and the Environment, 14*(2): 94–101. https://doi
    .org/10.1002/fee.1225

Appendix: Author's Comments on the Text

1   Changes in land use are the most important drivers, and of
    these urbanization and agriculture are the most significant,

according to the IPBES report from 2019. The world's
cities have doubled in size since 1992, and this growth is
expected to increase. Díaz, S., Settele, J., Brondízio, E. S.
(co-chairs) et al. (2019). *The global assessment report on biodiversity and ecosystem services: Summary for policymakers.*
Intergovernmental Science-Policy Platform on Biodiversity
and Ecosystem Services (IPBES). https://ipbes.net/sites/
default/files/inline/files/ipbes_global_assessment_report_
summary_for_policymakers.pdf

2   Slabbekoorn, H., & Ripmeester, E. A. P. (2008). Birdsong
and anthropogenic noise: Implications and applications for
conservation. *Molecular Ecology, 17*(1): 72–83. https://doi
.org/10.1111/j.1365-294X.2007.03487.x

3   Slabbekoorn, H., & Ripmeester, E. A. P. (2008). Birdsong
and anthropogenic noise: Implications and applications for
conservation.

Lightning Source UK Ltd.
Milton Keynes UK
UKHW010111111222
413712UK00005B/93